Fish Diseases and Health Management

Fish Diseases and Health Management

Ahtesham Malik

RANDOM PUBLICATIONS
NEW DELHI - 110 002 (INDIA)

Fish Diseases and Health Management

ISBN 978-93-51112-86-0

Published in 2014 in India by

RANDOM PUBLICATIONS

4376-A/4B, Gali Murari Lal, Ansari Road
New Delhi-110 002
Phone: +9111-43580356, 23289044
E-mail: randomexports@gmail.com; sales@randompublications.com; info@randompublications.com

Reprint 2021

Type Setting by: Friends Media, Delhi-110089
Digitally Printed at: Replika Press Pvt. Ltd.

Preface

Fish health management is a term used in aquaculture to describe management practices which are designed to prevent fish disease. Once fish get sick it can be difficult to salvage them. Successful fish health management begins with prevention of disease rather than treatment. Prevention of fish disease is accomplished through good water quality management, nutrition, and sanitation. Without this foundation it is impossible to prevent outbreaks of opportunistic diseases. The fish is constantly bathed in potential pathogens, including bacteria, fungi, and parasites. Even use of sterilization technology (i.e., ultraviolet sterilizers, ozonation) does not eliminate all potential pathogens from the environment. Suboptimal water quality, poor nutrition, or immune system suppression generally associated with stressful conditions allow these potential pathogens to cause disease. Medications used to treat these diseases provide a means of buying time for fish and enabling them to overcome opportunistic infections, but are no substitute for proper animal husbandry. Daily observation of fish behaviour and feeding activity allows early detection of problems when they do occur so that a diagnosis can be made before the majority of the population becomes sick. If treatment is indicated, it will be most successful if it is implemented early in the course of the disease while the fish are still in good shape.

Fish disease is a substantial source of monetary loss to aquaculturists. Production costs are increased by fish disease outbreaks because of the investment lost in dead fish, cost of treatment, and decreased growth during convalescence. In nature we are less aware of fish disease problems because sick animals are quickly removed from the population by predators. In addition, fish are much less crowded in natural systems than in captivity. Parasites and bacteria may be of minimal significance under natural conditions, but can cause substantial problems when animals are crowded and stressed under culture conditions. Disease is rarely a simple association between a pathogen and a host fish. Usually other circumstances must be present for active disease to develop in a population. The most obvious sign of sick fish is the presence of dead or dying animals. However, the careful observer can usually tell that fish are sick before they start dying because sick fish often stop feeding and may appear lethargic. Healthy fish should eat aggressively if fed at regularly scheduled times. Pond fish should not be visible except at feeding time. Fish that are observed hanging listlessly in shallow water, gasping at the surface, or rubbing against objects indicate something may be wrong. These behavioral abnormalities indicate that the fish are not feeling well or that something is irritating them. Low oxygen is a frequent cause of fish mortality in ponds, especially in the summer. High levels of ammonia are also commonly associated with disease outbreaks when fish are crowded in vats or tanks. Separate extension fact sheets are available that explain oxygen cycles, ammonia cycles, and management of these water quality problems. In general, check dissolved oxygen, ammonia, nitrite, and pH, during a minimum water quality screen associated with a fish disease outbreak.

This book will develop in the student an understanding of the principles of this subject. In order to make the subject simple and interesting every topic included in this book has been self-sufficient in itself and has been explained in the light of modern development in a simple and elegant style.

I thank all members of my team who have helped in the preparation of the book. My special thanks go to "Random Publications" who have published the book.

—Ahtesham Malik

Contents

1

Fish Diseases and Health Issues

By diagnosing a disease, illness, health or harassment problem a fish, coral as well as other invertebrate may be dealing with in your aquarium, here you can learn about how to quarantine an animal if necessary and take steps quickly to protect or treat them, whether using an ozone or UV sterilizer, medications, QT, or other means to care for them.

- Bacterial & Other Diseases (11)
- Coral Diseases & Health
- Disease Diagnostic Tools (5)
- Fish Euthanasia (7)
- HLLE / LLE Disease (11)
- Hyposalinity / OST (9)
- Ich Diseases & Parasites (21)
- Ozone & UV Sterilizers (8)
- QT Set Ups & Quarantining (5)
- Stray Voltage Problems (3)

Five Reasons that Marine Fish Die in Aquariums

In spite of what you may have heard from some "celebrities" in the press (P.E.T.A. and Snorkel Bob, on Maui in HI come to mind) statistically, fish will live longer in a well maintained marine aquarium than in the wild. Aquariums are a whole different story, though. A vast majority of salt aquariums are "closed systems" (not open to the ocean)...

Fish Diseases

By diagnosing a disease, illness, health or harassment problem a fish, coral as well as other invertebrate may be dealing with in your aquarium, here you can learn about how to treat fish disease by quarantining an animal if necessary and take steps quickly to protect or treat them, whether using an ozone or UV sterilizer, medications, QT, or other means to care for them.

Tips for Pre-Acclimating Saltwater Livestock

Tips for Pre-Acclimating Saltwater Livestock. There are steps you can take to help your new fish acclimate well to your tank even before you bring it home.

Mail Order Fish and Invertebrates

Important shipping and packing tips that should be considered before buying saltwater livestock on line or by mail order.

Acclimating Saltwater Fish the Drip Line Way

The pros and cons - instructions and tips for using the drip line acclimating method as a way to introduce new saltwater fish into an aquarium.

Lymphocystis Virus in Saltwater Aquariums

The appearance of the Lymphocystis virus in saltwater aquarium fish is fairly common. Much like the various forms of Ich, though not normally fatal in and of itself (death is usually caused by secondary bacterial or fungal infection), this virus can spread through an entire fish community in a very short period of time. Early diagnosis and treatment are essential while prevention is fairly simple.

Fish Disease Trouble Shooter

This is an easy to use trouble shooter from your Guides to quickly identify and obtain treatment options on the most common diseases of saltwater aquarium fish.

What Hyposalinity Kills

Osmotic Shock Therapy (also known as O.S.T. or Hyposalinity) is one of the most effective, non chemical treatments for Saltwater ich (cryptocaryon). Hyposalinity will kill corals and many other desirable animals in a reef tank, but can also be used to eliminate a number of SW aquarium pests.

Why Did My Fish Die?

Your fish suddenly died for no apparent reason? Here is a simple explanation of the causes of fish deaths in a saltwater aquarium.

How to Treat Ammonia Burns

How to Treat Ammonia Burns explains how to treat saltwater aquarium fish when they are exposed to an excess level of ammonia. Symptoms of ammonia burns include: cloudy eyes, frayed fins, rapid gilling and loss of appetite.

Coral Diseases and Health Issues

Information about coral diseases such as bacterial, white pox, band and rapid wasting disease, coral bleaching, shedding and other issues that pertain to the health of corals both in nature and in reef aquariums.

- Fish & Invert Care Topics
- Fish Diseases & Health

Coral Competition - Turf Wars in Coral Reef Tanks

Coral Competition - Turf Wars in hard and soft Coral Reef Tanks. Corals on the reef compete for space. So do the corals in your aquarium. Here is how it works and what you can do about it.

Why and How Leather Corals Shed

All Sarcophyton species leather corals periodically slough off a thin layer of tissue from their surface. Find out why they do this, and the view step-by-step coral shedding process pictures shown here.

Why Corals Change Color

What causes corals to fade and change color? Find out why it's primarily about light, and how this occurs.

Top Phosphate Removing Products

The accumulation of phosphate (PO4) in a reef aquarium effects the growth and health of corals. Here are some of the best products that are recommended for removing this harmful element in reef tanks.

Top Nitrate Removal Products

High nitrate is harmful to corals and other invertebrates in reef aquariums. Here are some of the best nitrate absorbing or removal products on the market that can help reduce this troublesome element in reef tanks.

What is Coral Bleaching, and Why Does it Occur?

From changes in ocean currents and temperatures, to sedimentation and pollution, the impact of these and other natural occuring events in nature that are suspected of causing coral bleaching on reefs around the world are those that effect reef tank corals. If you are having trouble with coral bleaching, take time to evaluate what situations...

Coral Disease Page

Even though The Coral Disease Page does not address how to treat coral diseases, but studies and researches them in nature, it is very helpful in identifying the many diseases that can afflict hard and soft corals from the many pictures they provide here.

Major Diseases of Reef-Building Corals

The NOAA's "CORIS" (Coral Reef Information System) site profiles 8 major diseases that can afflict reef-building corals in nature; black-band, dark-spots, red-band, white-band, white-plague, white pox and yellow-blotch disease, and coral bleaching. Photos are included.

Bacterial, Fungal, Viral and Other Fish Diseases

Articles and information to learn about diagnosing and treating bacterial, fungal, viral, and others diseases that can afflict fish in saltwater aquarium and reef tanks.

The Lymphocystis Virus in Saltwater Aquarium Fish

A virus which has been seen sporadically in the past seems to be popping up more and more on saltwater aquarium fish, lately. The virus been identified as the Lymphocystis Virus and presents as a white or beige colored cauliflower like growth or growths on the skin or fins.

Bacterial Diseases Profile

Learn how to identify and recognize the signs of both internal and external bacterial diseases or infections that can afflict saltwater fishes in aquariums, what causes this problem and how to properly treat it.

Popeye Causes and Treatments

Popeye is more a condition than it is a disease in saltwater fishes. From minor to severe cases, learn about why this problem occurs and how it can be treated.

Lymphocystis Viral Disease FAQs

In an Email discussion FAQ format, here are exchanges with Robert Fenner about Lymphocystis Viral Disease. Information to understand, diagnose, and treat this virus is found here, and the pictures can help identify this marine fish disease.

Lymphocystis Viral Disease Pictures

A couple of photos from Robert Fenner's WetWebMedia site of a *Heniochus acuminatus* (Poor Man's Moorish Idol) shown with the viral infection, Lymphocystis. This virus can appear on most any area of a fish's body as attached white tumors or lumpy growths.

Mycobacteriosis

From the AMDA "Medicine Chest", an article contributed by Lance Ichinotsubo about this disease, known as Piscene Tuberculosis in the past, that is doubtful to cure and that CAN be passed on to humans if caution is not used!

Virus & Bacterial Infections

From the Sea & Sky site, disease, treatment, and prevention information on Bacterial Finrot, Pseudomonas and Vibrio (Bacterial Infections) and Lymphocystis (Califlower Disease).

Other Fish Diseases

From the Sea & Sky website, information about Marine Fungus (Ichthyophonus or CNS Disease), HLLE, Poisoning, and Malnutrition.

A Report on PTSD

From the AMDA "Medicine Chest", an article contributed by By Morgan Lidster of Inland Aquatics that explains what the Post-Traumatic Shipping Disorder (PTSD) theory is all about, and how their observations seem to reflect that it largely effects imported Angelfishes, as well as a few other select species.

Marine Ich Diseases and Other Fish Parasites

White Spot, Black Spot, Velvet and Clownfish Diseases

By Stan & Debbie Hauter, About.com Guides

- hyposalinity
- quarantine tanks
- fish diseases

Learn all about how to diagnose and treat the many types of saltwater ich parasites that can plague and kill marine fish in aquariums, such as White Spot, Black Spot, Velvet and Clownfish Diseases, as well as other parasitic animals like fish flukes and lice.

Clownfish Disease - Brooklynella Diagnosis and Treatment

An infestation of *Brooklynella hostilis* is commonly known as Clownfish Disease, and although this type of ciliated protozoan saltwater ich is usually associated with Clownfishes as the name implies, it is still a parasitic organism can affect other species, most often being Angelfishes.

White Spot Disease, Saltwater Ich - Cryptocaryon Diagnosis and Treatment

White Spot Disease or Saltwater Ich is caused by an infestation of the ciliated protozoan *Cryptocaryon irritans*. Even though this ich organism progresses less rapidly than Oodinium and Brooklynella, in a closed aquarium system it can reach overwhelming and disastrous numbers just the same if it is not diagnosed and treated properly upon recognition.

The Life Cycle of Cryptocaryon Irritans

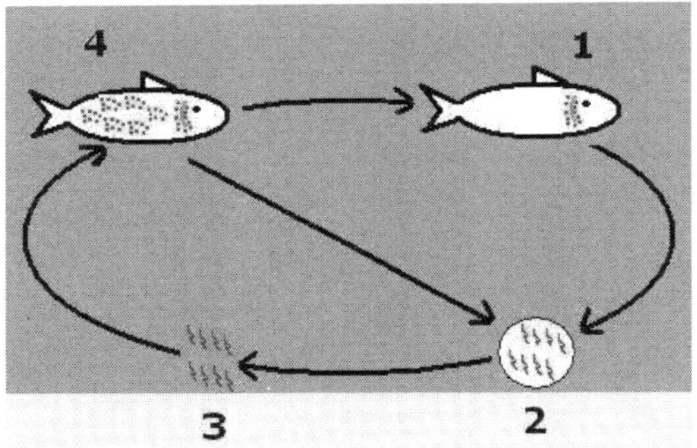

Figure: Stan Hauter

Cryptocaryon irritans is probably the most common type of ich outbreak encountered by saltwater aquarium keepers. The 4-stages of this organism's life cycle is simply outlined here, including a diagram, which will help you better understand the different phases of life of White Spot Disease, especially when it is most vulnerable to treatment.

Marine Velvet, Coral Fish Disease - Oodinium Diagnosis and Treatment

Marine Velvet or Coral Fish Disease is caused from an infestation of the single-celled dinoflagellate *Amyloodinium ocellatum*. Oodinium reproduces rapidly and can easily go undetected, as physical signs of its presence are typically not apparent until it's too late. It is not unusual for a run in with this type of saltwater ich to cause a total fish population wipeout, so knowing what sypmtoms to look for to diagnose it as soon as possible, as well as which treatments are most effective for irradication are critical.

New Marine Ich Wipes Fish Out in 24 - 48 Hours!

Resistant Cryptocarayon Shows Up In 06' - This strain of marine Ich is resistant to traditional treatments with Formaldehyde, Malachite Green, Copper and All the dyes. Greatly peaking my interest, I investigated further, and upon reading this next statement, my interest became concern.

Black Ich, Black Spot or Tang Disease

Black Ich or Black Spot Disease is actually an infestation of tiny *Paravortex* turbellarian flatworms, and even though the name Tang Disease implies it is ususally associated with Tangs and Surgeonfishes, it can effect other species. Although the worms are rather easy to get rid once they appear as black dots on fish, they can survive for several months without a host, and therefore can be difficult to eliminate from an aquarium.

Top Parasite Medications

Parasitic infestations in a saltwater aquarium are among the most feared and difficult to treat. There are a number of medications on the market designed to prevent and/or eliminate parasites. Some can be used with invertebrates, some can not.

Ways to Treat Ich Diseased Fish with Formalin

Over-the-counter formalin products typically made up of a 37% solution of formaldehype diluted with water are one of the most effective medications for treating various types of protozoan caused ich infestations (crytopcaryon and brooklynella), parasitic fish flukes, lice, worms (black ich), and fungal diseases. Find out how to use formalin in many different ways to safely treat diseased fish, such as by means of a quick dip, a timed bath, or long term in a QT.

Black Spot or Tang Disease

Black Ich Diagnosis and Treatment

By Stan & Debbie Hauter, About.com Guides

- ich diagnosis
- bacterial diseases
- hyposalinity

What is Black Spot Disease?

Typically a "disease" that is most commonly associated with Tangs and Surgeonfishes, but one that can be contracted by other species as well, what is referred to as Black Spot Disease, Tang Disease or Black Ich is actually caused by an infestation of tiny Genus *Paravortex* member turbellarian flatworms.

Although parasitic organisms that are much less dangerous and life-threatening, as well as rather easy to get rid of compared to other ich diseases such as Oodinium, Cryptocaryon, and Brooklynella, nonetheless it is a problem that needs to be treated upon recognition to eradicate infected fishes of these parasites.

The Life Cycle of These Worms

- Living in the substrate until mature, an adult worm seeks out a host fish.
- After feeding for about six days, it falls off into the substrate.
- About five days later the worm's body ruptures and releases a new population of young worms, and the cycle starts again, but in larger numbers.

Symptoms to Look For

Once these worms start feeding on a host fish they acquire pigmentation, which causes the appearance of black spots about the size of a grain of salt on the body and fin membranes. Worms that freely move about on fish, the spots do not always remain stationary. On light colored fish they are easy to see, whereas on dark colored ones they may go unnoticed at first. Fish will scratch up against objects in an attempt to dislodge the parasites, and if allowed to progress the fish become lethargic, loss of appetite and colors occur, rapid respiration develops, and secondary bacterial infections can invade damaged tissue sites.

Treatment Recommendation

Give all infected fish a freshwater dip, followed by a formalin bath and continue treatment in a QT.

Preventing Reinfestation

Reinfection will occur no matter how effectively the fish have been treated if these parasitic tubellarian worms are not eradicated from the main aquarium. Unfortunately because they can survive for several months without a host, this is often not an easy thing to do, but here are some suggestions.

- Leave the main aquarium devoid of all fish for several months.
- Young worms live in the substrate and feed on detritus and organic debris until such time they mature and go in search of a fish host. By siphoning up excess organic matter that can build up on the bottom of the aquarium can help to control their numbers.
- For fish-only tanks that have no freshwater sensitive invertebrates present, hyposalinity can be applied.

Clownfish Disease

Brooklynella Diagnosis and Treatment

By Stan & Debbie Hauter, About.com Guides

- saltwater ich
- fish diseases
- hyposalinity

Saltwater Aquariums Ads

- Goldfish Diseases
- Fibromyalgia Disease
- Liver Disease Symptoms
- Koi Disease
- Fish Disease

What is Brooklynella?

Brooklynella is a type of saltwater ich caused by an infestation of the ciliated protozoan *Brooklynella hostilis*. It is most closely and commonly associated with subfamily Clownfish members of the

Damselfish family, and therefore is typically referred to as Clownfish Disease. Although this parasitic scourge similar to others requires a fish host to survive, it is not particular in its quest to find one. Angelfishes, tangs or surgeonfishes, wrasses, jawfishes, and seahorses among others will host Brooklynella.

These protozoa reproduce asexually by means of simple binary fission through conjugation, which is why they are able to multiply so much more rapidly than Cryptocaryon (Marine Ich/White Spot Disease), and Oodinium (Velvet/Coral Fish Disease), and why it can kill fish within a few days and even hours upon recognition. For this reason accurate diagnosis and immediate treatment of all fish exposed to these life-threatening organisms is critical.

Symptoms to Look For

Most similar symptomatically to Oodinium, this too is a parasite that primarily attacks the gills first. At the onset fish may scrap up against objects, rapid respiration develops, and fish often gasp for air at the surface as the gills become clogged with mucus. Fish become lethargic, refuse to eat, and colors fade, but the most noticeable difference that sets Brooklynella apart from Oodinium is the heavy amount of slime that is produced. As the disease progresses a thick whitish mucus covers the body, usually starting at the head and spreading outward, skin lesions appear, and it is not uncommon for signs of secondary bacterial infections to arise.

Treatment Recommendation

Suggestions range from copper, malachite green and other remedies, with some recommended being used in conjunction with formaldehyde. However the general consensus is these types of medications are either largely ineffective or do not work at all, and that the best and most effective treatment for Brooklynella is formaldehyde alone. Typically a standard 37% formalin solution (shop & compare prices) is mixed with either fresh or saltwater in a separate treatment container, intitially all fish are given a quick dip or a prolonged bath, followed by continued treatment and care in a QT. Of course the longer fish are exposed to the formalin treatment, the more effective it will be at eliminating this “disease”. Whether to administer a dip or a bath to start with is something you will have to determine yourself, but there’s a very simple way to do this.

How to Treat Ich Diseased Fish with Formalin

If a formalin solution is not available for immediate use, temporary relief may be provided by giving fish a freshwater dip or bath. Even though this treatment will not cure the disease, it can help to remove some of the parasites, as well as reduce the amount of mucus in the gills to assist with respiration problems. Once the initial dip or bath is done, place the fish into a QT under hyposalinity treatment to help keep any possible new free-swimming protists from infecting the fish again, and then obtain a formalin medication as soon as possible to begin treatment.

Saltwater Aquarium Fish Disease Trouble Shooter

Saltwater Aquariums Ads

- Fish Tanks
- Fish Aquarium Supplies
- Sick Fish Treatment
- Saltwater Aquarium
- Fish Disease

One of the most frustrating experiences an inexperienced saltwater aquarium hobbyist can have is trying to diagnosis a disease in their tank.; The up side to the experience is that, once you have successfully diagnosed and treated a disease, the next time you see the symptoms it will be at an earlier stage, making a successful cure much more likely. The Fish Disease Trouble Shooter is designed to help you proceed through a step by step process of following a path of specific symptoms to identify the disease.; The Trouble Shooter isn't perfect as every situation is different and there are a number of variables which are difficult to screen.; However, taking a few minutes to run your situation through the Trouble Shooter should help you focus on fewer possibilities.

You will find a number of links to more information as you close in on the specific disease your fish probably suffers from.

Let's start off with some of the more obvious symptoms.

Fish and Coral Diseases

Diagnosing and Treating Fish and Coral Diseases

By diagnosing a disease, illness, health or harassment problem a fish, coral as well as other invertebrate may be dealing with in your aquarium, here you can learn about how to treat fish disease by quarantining an animal if necessary and take steps quickly to protect or treat them, whether using an ozone or UV sterilizer, medications, QT, or other means to care for them.

Marine Ich Diseases and Other Parasitical Fish Diseases

The many types of saltwater "Ich" parasites can plague a saltwater aquarium, killing the entire marine fish population in a matter of days. The parasites collectively known as "Ich" (or "Ick") can take many forms. Outbreaks of:

- White Spot (Cryptocaryon)
- Black Spot Disease
- Velvet or Coral Fish Disease (Oodinium) and

as well as a New Type of Marine Ich which can wipe out fish in 24 - 48 hours! and other parasitic animals like fish flukes and lice typically occur in a stable aquarium when new livestock is added, or when even one of the population becomes stressed. Early diagnosis and rapid treatment of the particular parasite strain are keys to an effective cure.

Neon Tetra Disease (Pleistophora hyphessobryconis)

Freshwater Aquariums Ads

- Sick Fish Treatment
- Fish
- Liver Disease Symptoms
- Disease
- Fish Aquarium Supplies

Disease Type: sporozoan

Organism: Pleistophora hyphessobryconis

Names: Neon Tetra Disease, Pleistophora

Description: Neon Tetra disease is more common than many aquarium enthusiasts realize, and affects species beyond neon tetras. Named after the fish that it was first identified in, the disease strikes members of the tetra family most often. However, other popular families of aquarium fish are not immune. Cichlids such as Angelfish, and Cyprinids such as Rasboras and Barbs, also fall victim to the disease. Even the common Goldfish can become infected. Interestingly enough, Cardinal tetras are resistant to the ravages of Neon Tetra disease. Caused by the sporozoan, Pleistophora hyphessobryconis, the disease is known for its rapid and high mortality rate among neons. To date there is no known cure, the only 'treatment' being the immediate removal of diseased fish to preserve the remaining fish.

The disease cycle begins when parasitic spores enter the fish after it consumes infected material, such as the bodies of dead fish, or live food such as tubifex, which may serve as intermediate hosts. Once in the intestinal tract, the newly hatched embryos burrow through the intestinal wall and produce cysts within the muscle tissue. Muscles bearing the cysts begin to die, and the necrotic tissue becomes pale, eventually turning white in color.

Symptoms:

- Restlessness
- Fish begins to lose coloration
- As cysts develop, body may become lumpy
- Fish has difficulty swimming
- In advanced cases spine may become curved
- Secondary infections such as fin rot and bloating

During the initial stages, the only symptom may be restlessness, particularly at night. Often the first thing an owner will notice is that the affected fish no longer school with the others. Eventually swimming becomes more erratic, and it becomes quite obvious that the fish is not well. As the disease progresses, affected muscle tissue begins to turn white, generally starting within the color band and areas along the spine. As additional muscle tissue is affected, the pale coloration expands. Damage to the muscles can cause curvature or deformation of the spine, which may cause the fish to have difficulty in swimming. It is not unusual for the body of the fish to have a lumpy appearance as the cysts deform the muscles.

Rotting of the fins, especially the caudal fin, is not uncommon. However, this is due to secondary infection rather than a direct result of the disease itself. Bloating is another secondary infection.

Treatment

- *None, separate or euthanize diseased fish:* There is no known cure. To ensure all fish are not lost, remove diseased fish from the tank. Some species, such as Angelfish, may live for quite some time. However, they should be separated from uninfected fish to avoid spreading the disease.

Prevention:

- Quarantine new fish for two weeks
- Maintain high water quality
- Do not purchase from a tank with ill fish

The best prevention is to avoid purchasing sick fish, and to maintain high water. Carefully observe the suppliers fish. Do not purchase any fish from tanks where there are sick, dying, or dead fish present. Fish that do not school, or hang apart from the others, should be suspect.

Bacteria and Antibiotics

Like Little Chemical Factories: Bacteria are minute organisms invisible to the naked eye. There are many different types of bacteria and they are probably the most diverse group of organisms on the planet – each like a little chemical factory. Very few species are pathogenic (disease-causing). Those that are harmful can be divided into two basic types; Gram-negative and Gram-positive. This simple grouping is based on a staining technique in which Gram-negative bacteria stain red, while the Gram-positives stain blue because of differences in the cell-wall structure.

Bacteria Stained using Gram Differential Staining Method

Most bacterial fish pathogens, such as *Aeromonas, Pseudomonas, Vibrio, Flavobacterium* and *Cytophaga* are Gram-negative bacteria. These are the bacteria that are usually involved with bacterial disease such as ulcers, fin rot, acute septicaemia and bacterial gill disease. Less common pathogens are *Mycobacterium* and *Norcardia sp.* which cause chronic granulomas (or abscesses).

How do Bacteria Cause Disease?

Although they are incredible small most pathogenic bacteria have tremendous reproductive potential. They simple divide in two and once each half has re-grown sufficiently they divide again. Under ideal conditions the reproductive cycle can be as little as 20 - 30 minutes, which means that just one bacterium can multiple to several million within 24-hours or so!

As I have said, bacteria are like little chemical factories. Some bacteria produce toxins that are excreted into the blood and tissues of the host. Other bacteria, particularly the Gram-negatives do not secrete a soluble toxin but make an endotoxin that is liberated when the cell dies and disintegrates. These endotoxins are usually lipopolysaccharide structural components of the bacterial cell-wall (specifically the lipid A portion). In addition to toxins the virulence of many bacteria is partly due to the production of extracellular enzymes, which attack healthy fish cells.

So although tiny, the net effect of millions of bacteria can quickly overwhelm the defences of the fish host – which is why early treatment is so vital if the fish is to survive.

What Antibiotics are Available?

Every country has its own regulations on the control of antibiotics and therefore availability will vary. The range of drugs available or licensed for use with fish is generally small. In the UK all antibiotics are classed as POMs (Prescription only Medicines) and are only available under veterinary supervision. Among the antibiotics in common use are;

- Broad-spectrum penicillins that include amoxycillin and ampicillin. These are bactericidal in action and are effective against Gram-positive and Gram-negative bacteria, but not very effective against *Pseudomonas*.
- Chloramphenicol is a broad spectrum antibiotic, but again it is not usually effective against *Pseudomonas*. Because it is used in human medicine it is illegal for use in food animals in most countries. Some people are sensitive to the drug so rubber gloves should be worn when it is handled.
- There are several potentiated sulphonamide drugs in which a sulphurdrug is combined with trimethoprim. They are broad spectrum and mainly bacteriostatic. They are well absorbed from the gut when used with medicated food but poorly absorbed from water. They can sometimes form crystals in urine and should not be used where there is already kidney damage. They should not be used at the same time as organophosphates. Co-trimazine is a combination of trimethoprim and sulphadiazine and Borgal is a combination of trimethoprim and sulfadoxine. They are effective against *Aeromonas* and other Gram-negatives, but not particularly effective against *Pseudomonas*.
- Gentamicin is a broad spectrum antibiotic with a bactericidal effect. It is useful against all Gram-negatives including *Pseudomonas* but can be toxic to kidneys. There is concern that bacteria may acquire rapid resistance, so it would not be the drug of first choice but does have a use when other drugs do not work.
- Nitrofurans are a group of synthetic antimicrobials. Both nifurpirinol and nitrofurazone are effective against many fish pathogens. They are well absorbed through the skin, making

them ideal for bath treatments. However, they are carcinogenic and mutagenic and their use with food fish is illegal in most countries.

- Oxytetracycline is bacteriostatic drug with a broad spectrum of activity. However, there is widespread resistance and it has little effect on *Pseudomonas*. The long-acting injectable formulation has caused sterile cysts at the injection site. When used for bath treatments it readily chelates calcium and magnesium ions found in hard water, significantly reducing its effectiveness. It is also light-sensitive when used as a bath treatment, turning brown as it decomposes. The degraded form can be toxic to both fish and humans. Some studies show that oxytetracycline is immunosuppressive in some fish species.
- The quinolones are narrow spectrum range of drugs acting mainly against Gram-negatives such as *Pseudomonas* and *Aeromonas*. The first generation quinolones include oxolinic acid to which there is now widespread resistance. Newer quinolones include enrofloxacin and sarafloxacin. These are effective against *Pseudomonas* and *Aeromonas* and as yet there is little resistance. All quinolones are chelated in hard water, so they are not practical for bath treatments. They are well absorbed through the gut, making them useful for medicated food.

Antibiotic Sensitivity

Bacterial resistance is an issue that needs to be considered when choosing an appropriate antibiotic. While it may be necessary to start a course of treatment based on personal experience, it is also prudent to have bacterial identification and antibiotic sensitivity tests carried out at the same time. At worse this simply confirms the treatment choice, while possible saving valuable time and money if the initial course of action is ineffective. A sterile swab is used to take a bacterial sample from either the posterior kidney or other organ of a newly dead fish, or alternatively from a body ulcer. When taking a swab of a body ulcer it should be sampled from the infected rim of the ulcer only.

The swab is then sent to a specialist laboratory for analysis and testing. Your local veterinarian can arrange this. The results from a body ulcer can be difficult to interpret because of the problem of secondary pathogens. It is quite common for samples from body lesions to show a mixed growth of several bacterial species and takes some experience to determine the relevance of the results.

Fish Diseases

Fish disease is a fact of life and in common with all pet owners, fish-keepers need to be able to recognize the signs of ill-health in their fishy charges. These pages give a comprehensive view of common diseases including typical signs of disease and how to systematically diagnosis both health problems and their causes.

An Overview of Aquatic Health: explains why disease occurs. It explains the various diagnostic steps to take and outlines the various treatment methods.

The Fish Health Work-up: an invaluable step by step guide to diagnosis health problems and their causes.

Clinical Signs of Disease: Although it is impossible to make a diagnosis based purely on behaviour we can use typical behaviour and physical changes as indicators of possible health problems.

Argulus: The fish louse. A real nasty. In the confines of a pond or tank these can be real killers and finding even one louse warrants immediate treatment.

Bacterial Infections: including fin rot and ulcers are a serious threat to health.

Bacterial Gill Disease: is usually a 'management' problem, arising from less than ideal conditions.

Chilodonella: a common parasite, often occurring at low temperatures. It is a threat as it often causes severe damage before any signs are evident.

Costia: is able to reproduce rapidly under ideal conditions and cause serious tissue damage, particularly in the gills.

Dermocystidium: a quite horrifying condition that leaves koi covered in angry-looking lesions - but then disappears literally overnight!

Fin rot: With few exceptions, fin rot is always precipitated by stress and is often one of the first signs that a health problem exists.

Skin and Gill Flukes: are common parasites that are often the precursor to gill damage, bacterial infections and ulcers. Treatment can sometimes be tricky.

Fungus: *Saprolegnia* infections can spread rapidly and even though the damage is often superficial, prompt treatment is essential to avoid fatalities.

Parasites - an overview: Fish are potential hosts to a range of parasites. While small numbers of parasites do little harm in most cases, in large numbers they can seriously compromise the health of the fish

Trichodina: a common parasite. At low levels it doesn't pose a health problem. However, as with most parasites they have tremendous reproductive potential.

Bacterial Body Ulcers: are a common problem. An early, accurate diagnosis and treatment is vital if losses and expensive treatment costs are to be avoided.

White Spot: a nasty parasite capable of causing serious tissue damage. Up to 100% fatalities can occur without prompt treatment.

I should stress, that although these diseases sound daunting, most are easily cured if treated early enough.

The key is observing behaviour. Any changes from normal behaviour that persist for more than a day or two must be investigated - as it usually indicates something is wrong.

However It is important to understand that a definite diagnosis can't be made purely on the basis of the behaviour changes - a proper examination is usually required.

2

Fish Diseases: Diagnosis and Possible Cures

As usual, avoiding problems is always better than fixing them. All in all, the less stress put on your fish and the better your maintenance regime, the better your chances of having happy, healthy, long-lived fish. Keep all equipment clean, do not over-clean filters, do not over-populate, feed sparingly, perform good quality water changes regularly, etc etc etc. But, if your fish do become infected with a disease after all, here's a list to help you identify it and and try a cure:

White Spot (Ichthyophthirius)

A parasite that we will all encounter whilst keeping fish. As an adult, it is embedded in the skin of the fish causing irritation. Your fish will probably be glancing off rocks and plants to alleviate the itching. The parasite will feed and grow on the blood and skin cells of its host for a few days, until it is fully grown.

It then bores its way out of the skin and drops off into the substrate. It then forms a cyst which goes through rapid cell division until about 1000 young are released into the water to start the whole process again.

The whole process takes just five days at 27°C. There are cases where the parasite actually lies dormant in the skin of the fish and will not emerge until it is ready, making eradication quite difficult. The best time to kill them is when they are free swimming and looking for a host. Treatment must be external and aimed at the free swimming stage, hence the need to dose again after a few days. Suitable cures are widely available.

Velvet (Oodinium)

Velvet is caused by a dinoflagellate, classified by some as an alga because it carries Chlorophyll. There are two main species, *Oodinium limmneticum* and *Oodinium pillularis*. The free swimming stage of Velvet settles on the skin and gills of the fish, adhering at first by its long flagellum, later putting out pseudopodia, (similar to fingers) that penetrate the skin and give it a very firm grip. Velvet has a similar life cycle to White spot: feeding and becoming a cyst, it produces upward of 200 young before dropping off. The appearance is similar to gold or brown dust over the body and fins which at times may appear to move. Symptoms are similar to White spot, glancing off rocks, etc. As Velvet is highly contagious, it is important to eradicate this problem as soon as possible. Treatment is aimed at the free swimming stage and there are good cures available from your local store. Copper sulphate can be used at a concentration of 0.2 mg/L or ppm. This should be repeated after 3 days to ensure eradication.

Fungus (Saprolegnia or Achyla)

Fungi are in fact colourless plants, they are very diverse in form and have not been studied that much in fish. Saprolegnia and Achyla are genera of fungi that attack weakened and injured fish, usually settling on damaged skin or gills. They will also attack the eggs of fish. The spores enter the aquarium from the air so there is no way to stop possible outbreaks. Threads of fungus spread under the skin forming a web like structure and eventually produce tufts of external hyphae that may be large enough to look like cotton wool. These form spores that become free swimming and can then go on to infect other fish that are weak or injured. It is important that you remove any fish you suspect immediately to a hospital tank for treatment. Symptoms of fungus are a grey or whitish growth in the skin of the fish, often associated with visible damage. Treatment can be with Malachite green in a separate tank for 30 seconds at a strength of 60 mg/L. Repeat treatments may be necessary. The fungus is stained by the Malachite and usually drops off within a few hours. Keep the fish under observation for 2 or 3 days as fungus can sometimes regrow.

Mouth Fungus (Columnaris)

Although Columnaris resembles fungus it is in fact a Bacterium. It is usually found just around the mouth area, rarely spreading. The first signs are a thin white line around the lips, which then grows into white or grey short tufts that resemble fungus. Early treatment is needed if you are going to save the fish's life, due to the toxins released and the fact that the fish cannot feed. The best and most effective cure is to use Anti-Biotic. Penicillin is very effective at 10.000 units/L, with a second dose after two days. Remember to remove all filtration whilst dosing to avoid killing the beneficial bacteria.

Neon Disease (Plistophora)

Named after the fish that it was first identified in, but this problem can affect all other tetras and it is now known to affect other species. The organisms lie in the muscle tissue in the form of cysts which burst and release spores, the spores then bore deeper in to the muscle and then repeat the process. Eventually some of the spores will reach the water, probably via the gut. These then go on to infect more fish, usually by ingestion. I personally think the best way to control this is to humanely kill the infected fish as there is no reliable cure.

Digenetic Flukes (Metacercaria)

Typically seen in newly imported fish and in two forms. Black spot is caught from snails that release the Cercaria. Larval forms penetrate the skin and encyst in the tissue and may be seen as red or black nodules. If eaten by a Bird for instance they then develop into adults. Sanguinicola disease passes from fish to snail and then back as minute worms that live in the fishes bloodstream. There they lay eggs that block up the blood vessels which in turn causes Necrosis. As far as I know there is still no cure for this problem.

Tapeworm Larvae (Cestoda)

Not commonly seen in fish, but can be introduced in tubificid worms as food. They then become adults in the gut of the fish, usually about 3mm in length. Not seen very often in the aquarium, but if seen there are commercial cures available, best to consult a Vet.

Costia

Costia necatrix is rather rare in aquariums, but if seen as skin cloudiness is easily eradicated by raising the temperature to 32°C, otherwise treat as Velvet.

Chilodonella, Trichodina

These actually sit on the skin and are more easily eradicated. Usually spotted by the fish glancing off objects, clamped fins and gasping at the surface as the gills are often infected. Treat as Velvet.

Red Pest (Bacterial)

Red pest is so called because of the red streaks that occur on the skin, that may lead to ulcerations, and fin or tail rot, causing parts of the fins or tail to actually drop off. This bacteria infects the fish internally and externally, and treatment is not usually effective. If fish are only lightly infected you can try to eradicate this problem by sterilising the aquarium with Acriflavine or Monacrin. Giving the aquarium a good clean up is also helpful, feed lightly whilst treating. Personally I would not bother with treatment as by the time you notice this bacteria it is usually well advanced, and the fish are already suffering.

Fin Rot (Bacterial)

Usually the secondary stage of Red Pest (above). Several bacteria can cause it. If the fish reaches this stage then you will have to use Anti-Biotic to eradicate. Remember to remove all filtration as the Anti-Biotic will kill the beneficial bacteria. Typical symptoms are: discoloration of the fins (in early stages may go unnoticed), fins become ragged and frayed, and in terminal stages only shreds of the fin or tail remain. Can lead to secondary fungus infections. The main causes are improper maintenance routine, infrequent water changes, temperatures too low for the fish being kept. This is a highly contagious disease.

Ichthyosporidium

This fungal disease is widespread, attacking the liver, kidneys and pretty much everywhere else. Infection begins via food and the parasites infect the blood stream, settling down as brown cysts. These then grow and produce daughter cysts by budding or by the mother cysts bursting and releasing the young. The symptoms depend on what part of the body is infected. Fish may become sluggish, show hollow bellies, lose balance, and eventually external cysts and sores. By this time it is usually to late to save the fish. Treatment is difficult, but treat as for Saprolegnia.

Pop-Eye (Exophthalmia)

Pop-eye may not be caused by infection and there are various causes. Easily identified by the protrusion of one or both eyes. It is commonly caused by excess gas in the system, brought about by super-saturation of gas in high pressure water mains. It can also be associated with Dropsy or Ichthyosporidium. Look for bubbles of gas in the eyes, this indicates the bends, as does nervous upset, distress or just odd movements. If these signs are obvious then lower the temperature slowly to increase solubility of the gas (usually nitrogen) reduce aeration and wait to see if there is any improvement. In marine aquaria, copper poisoning can be one of the causes. Pop-eye can also be caused by hormonal imbalance, for which there is no cure.

Hexamita (Hole in the Head)

Usually associated with Discus, this Flagellate will in fact affect other fish, particularly large cichlids. Hexamita infects fish by ingestion of food and begins life in the gut. Infected fish will usually appear thin in the stomach and may have sores or ulcers around the head. A bad case looks as if someone has chiselled out holes in the head. It is said by some that this will cure itself by generally cleaning the aquarium and paying extra attention to cleanliness and good quality water changes. There are cures available from your Local fish store.

Argulus, Ergasilus, etc

Fish lice (Argulus) are in fact a small crustacean that attaches itself to its host and then proceeds to feed on its blood. It is thought that they can transmit disease, so should be removed quickly. On large fish this can be done with forceps, unless you have a mass infestation. Potassium permanganate in a hospital tank at 10 mg/L for 10 to 30 minutes is effective. Or, alternatively, you can use insecticides but I am not sure about the use of insecticides in the aquarium. I personally would try a salt bath first to see if they drop off. Ergasilus is also a crustacean that affects the gills of fish. Try the above cures or ask for help.

Gyrodactylus Monogenetic Flukes

In fact a flat worm that in the early stages can be mistaken for white spot. They infect the gills and skin of their host and produce live young. Symptoms are pale skin, drooping fins, rapid respiration and even emaciation. Treat as for Argulus, Potassium Permanganate or alternatively use Formalin, but only in a hospital tank. Use 2 ml of 40% solution per 10 litres for 45 minutes. Salt bath at 15 g/L for 20 minutes may also work.

Nematodes

These threadworms will infect all parts of the body, but are only seen when hanging from the anus of the infected fish. Consult a specialist, or personally I would consult a Vet. There are theories that threadworm treatment for cats can be used mixed in with food, but I do not know the amounts needed.

Lymphocystsis

A viral disease that causes the host cells to swell up giving rise to tumours, that almost look like spawn (granular) attached to the body and fins. There is no guaranteed cure so it is probably best to euthanize the fish.

Glugea, Henneguya

Sporozoans that form large cysts that look similar to Lymphocystsis and is also incurable. Dispose of the fish humanely.

Tuberculosis

Symptoms are Hollow belly, Knife back, Pale skin, Bending of spine. They also lose their appetite which does not help them. If suspected, it is best to dispose of the fish immediately. Tuberculosis can be passed on to humans, but only if you have open wounds on your hands when placing them in the aquarium.

Trynoplasma

Sleeping sickness infects the blood just as it does in humans. Infected fish are drowsy, may swim strangely and become thin. It is usually seen in pond fish that have been recently moved into an aquarium. I think it is brought on by warming up the fish too quickly. There is no cure.

Dropsy (Aeromonas)

Dropsy usually takes one of two forms: body swells due to fluid accumulation, which causes scale protrusion, and also protrusion of the scales without the body swelling up - false dropsy. Dropsy is caused by a bacterial infection of the kidneys and other internal organs, causing fluid accumulation or even renal failure. Treatments are varied, adding salt at a rate of 20 mg/L seems to help. With modern Anti-Biotic this problem can be cured.

Most Common Fish Diseases

Unfortunately, even the most dedicated aquarist will from time to time have an outbreak of disease in the aquariums. How you handle these situations may be critical to your success as a hobbyist. One of the most common mistakes aquarists make is to medicate too soon. Very often the fish will fight off the disease on its own, provided that its immunity system is strong. I rarely medicate an aquarium. Rather, I rely on a series of partial water changes to rid the aquarium of parasites, pathogens, and other disease causing organisms. What follows are some of the diseases you are likely to run into and my recommended treatment. Please be aware that this is from my personal experience.

Ich (White Spot)

This is a parasite which resembles grains of salt on your fish. At first the white spots are small, usually present within the gills of the fish. Your fish may breath heavily or scratch against decorations in an attempt to remove the parasite. Within a day or two the parasites will have grown and multiplied, covering the body of the fish with tiny white spots. There are many techniques available to fight off Ich. The most common involves using copper sulfate, which kills free swimming parasites. Copper is very effective but can be stressful for your fish, especially scaleless fish. Copper can also have a negative effect on your biological filter bed, if your tank is not very well established. It is also not healthy for your live plants. A good alternative is to use Malachite Green. Malachite Green is commonly sold in pet stores, usually in combination with Formalin. Some common medications include Noxich and QuickCure. I have had excellent luck with both of these products. You should still use precaution when treating tetras or scaleless fish, although these products are safe with Live Plants. The usual recommendation is to treat at 1/2 dose for sensitive fish. Another method for treating Ich which can be useful is to slowly raise the temperature of your aquarium to a little above 29°C. This will dramatically speed up the life cycle of the parasite, driving it out of the system. However, this method relies on your fishes immunity system to fight the disease. Many hobbyists also advocate the use of salt to fight off Ich. They recommend adding 1 teaspoon of salt to 4 liters (1 gallon) of water. This will effectively irritate the skin of the fish, causing them to produce and excessive amount of slime. The slime coat serves as a protective layer to fight off parasites. I personally do not believe in this technique or any technique which causes unnecessary stress on the fish. There are other techniques available which are just as effective.

Velvet

This is also a parasite, but is luckily not as common as Ich. Velvet appears as a rust colored parasite which is very small and often difficult to see. It is usually clustered very heavily on the fish, forming sort of a rust colored blanket. Velvet attacks your fish very quickly and could even wipe out a tank in a matter of 2 days. To be honest, I have not come across an effective treatment. If I had an outbreak I would attempt adding salt and using copper in combination with Malachite Green. I would also raise the temperature and do a 90% water change.

Fungal Infections

A fungal infection appears as a white cotton growth on the body or fins of the fish. The majority of white spots you see on your fish are going to be parasites. In the slim chance that you have a fungal infection then I would recommend not treating at all. A true fungal infection is practically harmless to your fish and should be attacked successfully by your fish's immunity system.

Bacterial Infections

These usually appear as red patches or streaks on the body of the fish, or sometimes appear as damage to the fins of the fish, such as "fin rot". Bacterial infections are almost always due to poor, declining water conditions. This does not necessarily have to be high levels of ammonia, nitrite, or nitrate. Lack of regular water changes will result in increased numbers of pathogens and other disease-causing organisms which thrive in neglected aquariums. Bacterial infections are also commonly associated with parasite infection, as a secondary infection. I rarely recommend medicating for a bacterial infection as a series of partial water changes can usually eliminate the problem. Medicating can also have an adverse effect on your biological filter, as medications will not distinguish the good bacteria from the bad.

Prevent Rather Than Treat!

All in all, the majority of infections can be eliminated or prevented by frequent partial water changes, and ensuring that your fish lives in an appropriate environment (tank size, tank mates, etc) and water quality. Medications are rarely needed and should be avoided whenever possible.

Note: recent posts on Aqualink have revealed that Malachite Green increases in toxicity to fish as the temperature increases. You may want to reconsider your decision to use Malachite Green if you intend to raise the temperature at the same time, or if you already maintain your temperature at a higher level than normal. For the record, I have not personally experienced this problem.

Using Heat to Treat Ich in Freshwater Tropical Fish

What is Ich?

Ichthyophthirius multifilis (Ich) is a very common protozoan parasite that finds the aquarium environment a most hospitable location for carrying out its life cycle. Because of the low densities of fish in large volumes of water in the wild, even though every cyst can release hundreds to a thousand or so swarmers, few if any will manage to find a suitable host. Those fish that do get infected will likely have only a single or very few parasites attached to its body, which will soon fall off. The fish will heal and be none the worse for it. In the wild, Ich is really no big deal. A home aquarium, by contrast, is a confined space packed with many fish (relatively speaking). A single Ich parasite brought in on a new fish can quickly multiply and decimate an entire tank full of fish. Unlike in the wild, every larva has an excellent chance of finding a host, and each fish can end up with hundreds of parasites attached to its gills, fins and body. Such an onslaught is too much, and the fish dies from respiratory distress caused by gill damage, osmoregulatory imbalance caused by the many punctures in the fish's skin, or secondary infections of bacteria or fungi that attack the open wounds.

Ich is also known as White Spot Disease. It is characterized by the presence of small white spots, like a sprinkling of salt, that may appear anywhere on the fish. The fish may dart about in the water, and flick or rub its body against the substrate or other objects in the tank. As the condition advances, the fish may become very slimy, clamp its fins, stop eating, and become very listless. The standard chemical treatments are often effective, but they can be just as toxic to fish and plants as they are to Ich. Sometimes the medications are not effective. Many aquarists are interested in trying a more natural approach that doesn't subject their fish (or themselves) to potentially dangerous chemicals.

Ich Life Cycle

To better understand how it can be treated, a basic understanding of Ich is helpful:

- Parasitic Stage — The first stage (trophozoite) is embedded in the skin of the fish, and feeds on fluids and tissue cells. The fish tries to protect itself by producing more cells around the trophozoite, which leads to the formation of the tiny white spots.

- Intermediate Stage — Eventually the trophozoite morphs into a trophont, detaches from the host leaving an open wound, and swims freely until it locates a suitable place to settle.
- Reproductive Stage — Once the trophont settles, it forms a sticky wall around itself and becomes a cyst (tomont). Inside, the tomont divides into many hundreds or even thousands of tomites.
- Infectious Stage — The tomites are released into the water column and swim freely until they attach to a host, starting the cycle over again. If a tomite doesn't find a host within a short period of time, it will die.

The Ich life cycle is temperature dependent. Higher temperatures within its livable range speed up every stage of the life cycle, while the lower temperatures will slow it down. At 18°C/64°F the cycle takes 10-12 days to complete.

It has been found that Ich does not infect new fish at 29.4°C/85°F (Johnson, 1976), stops reproducing at 30°C/86°F (Dr. Nick St. Erne, DVM, pers. comm.), and dies at 32°C/89.5°F (Meyer, 1984).

Treatment

Now that we know a little more about Ich, we can develop a safe and effective natural treatment plan to eradicate it. A multi-pronged treatment plan offers the most assurance of complete eradication of Ich and TLC for the fishes in your aquarium. As with any treatment, carefully observe the reaction of your fish to any changes you make in their environment. If an adverse reaction occurs, discontinue and try another approach.

- Increase temperature to 30°C/86°F. With tropical fish, an increase in temperature to 30°C/86°F is usually very well-tolerated. Since this temperature prevents reproduction of Ich, it can theoretically cure the problem by itself. So the first step would be to increase the temperature slowly, 1°C/2°F per hour until the correct temperature is reached. This temperature should be maintained for 10 days, and then slowly returned to normal. Some fish can tolerate higher temperatures. If your fish are more heat tolerant, try increasing the temperature to 32°C/89.5°F for the first 3-4 days to kill the Ich. Then reduce temperature slowly to 30°C/86°F, and hold it there for an additional 6-7 days, or until a total of 10 days have passed. Gauge the heat tolerance of your fish by observing their reaction.

- Increase aeration. Increased temperature leads to increased metabolism, which enhances the fish's immune response but also increases oxygen demand. Oxygen is lower in warmer water, so it is very important to increase surface agitation during the treatment to increase oxygenation. In planted tanks with CO_2 injection, the CO_2 should be turned off and extra aeration should be provided. Carefully observe your fish, watching for signs that they are not getting enough oxygen. If fish are gasping at the surface, you need to provide more aeration. Aeration can be increased by reducing the water level so the filter return makes more of a waterfall and splash, and/or use an airstone placed close to the surface of the water.
- Do daily partial water changes. 25% daily partial water changes will provide several benefits: It will keep the water very clean, which will help fish cope with the stress of the disease. It will remove some of the trophonts and tomites. It will add oxygen. This author also recommends the use of NovAqua+ to condition the change water. This product is a dechlorinator and has several additional benefits that help fish under stress, including sealing of the wounds caused by the Ich. If the water changes seem to stress the fish, reduce the size and/or frequency of the water changes.
- Use a Micron Filter. The Aqua Clear Quick Filter used with a power head is an easy and inexpensive way to capture both free-swimming stages and the cysts of Ich in water that passes through the filter. A diatom filter can also be used. Both of these filters trap particles as small as one micron in size. The smallest stage of Ich, the free-swimming, swarming tomite, is approximately 30 microns, large enough to be trapped in this type of filter. Change the filter daily with the water changes. The Quick Filter cartridge can be cleaned and reused. Rinse thoroughly in very hot water, or boil for a few minutes to kill any stage of Ich that may be trapped inside. Or use a fresh cartridge. Make sure your fish are comfortable with the current caused by the additional filter.

The following optional procedures where appropriate are also beneficial:

- Remove Gravel. In a non-planted aquarium and where practical, the temporary removal of the gravel reduces attachment sites for the tomont and makes it easier to siphon the floor of the aquarium where many tomonts will be located.

- Use salt. In a non-planted aquarium with tolerant fish, the addition of Aquarium salt at the rate of 1 teaspoon per 4 liters/ 1 gallon of water disrupts the fluid regulation of Ich. Do not add salt crystals directly to tank. Always dissolve salt in a small amount of tank water before adding to tank. This dosage may be repeated every 12 hours for a total of three treatments. When Ich is gone, salt is removed with daily 25% water changes.

A separate salt bath is very effective for individual fishes. The higher concentration of salt will destroy embedded trophozoites on the body of the fish by preventing them from maintaining fluid balance. To prepare a salt bath for small-medium size fish, dissolve 2 Tablespoons Aquarium Salt in 4L/1 Gal. of conditioned tap water that is the same temperature as the tank water. Stir well to make sure all salt is dissolved before using.

Place affected fish in a container with enough tank water to cover the fish with a little room to spare. A wider container is better than a tall, narrow one, as it allows for better oxygenation of the water, and room for the fish to move about. Add salt solution very gradually until fish just starts to show mild stress. You may not need to add all of the solution. Leave in bath for 30 minutes, carefully observing the fish's reaction to the treatment. Return the fish to fresh water immediately if it rolls over or tries to jump out. Observe the fish carefully during the entire bath period. After the 30 minutes, dilute the salt bath in stages with tank water. Gently net out fish and return it to the main tank. This procedure can be repeated after 48 hours, up to 3 treatments. Do not use salt in planted tanks or with sensitive, soft water fishes. Do not pour bath water into main tank. By the end of the 10 days of treatment, all Ich should be gone and the fish will have healed completely from the effects. Keep the temperature up for the full 10 days to ensure that the Ich is gone.

Most strains of Ich should be eradicated by this heat treatment, but bear in mind that there is at least one strain in Florida that is known to be heat-tolerant. If you do not notice a marked improvement after 3-4 days of this treatment, you should stop and consider another approach.

Prevention

It is commonly thought that Ich lies dormant in all fish tanks, just waiting for an opportunity to attack. This is not accurate. However, it is possible that a resistant fish or group of fish can harbor a low level of Ich infection without obvious symptoms until stress of some kind reduces the fish's ability to fight it off. Then the infection becomes full-blown, seemingly out of nowhere. While this is a possibility, the most common and most likely cause of ich in any tank is through the introduction of new fish.

Ich is present in many fish bred in fish farms and distributed to pet stores. Carefully inspect each new specimen you are considering for purchase. Even fish that appear perfectly healthy can still harbor ich in the gills, or have a new infection that doesn't yet show the typical symptoms. If possible, quarantine all new fish in a separate tank to prevent an infestation of Ich, as well as other potential parasites and diseases. The quarantine period should be two weeks or longer, during which time you should carefully observe the new fish. Anti-parasitic medications or foods can be used during this time to prevent the spread of certain other types of parasites, such as worms and flukes.

3

Fish Disease Treatments

Treatments

Fish disease treatments should be straightforward provided that the problem has been accurately diagnosed at an early stage. However, it is very important to bear in mind that many fish are killed every year by the improper use of medications.

Disease Treatments: an overview of how treatments work, why they often don't work and why they sometimes kill fish.

Disease Treatment Basics: When you have fish health problems, this guide details the steps to take, and what to do if it isn't possible to make a definitive diagnosis.

Antibiotics: are an important treatment option against bacterial diseases if they are used properly.

Injecting Antibiotics: gives the best results against bacterial infections - especially when dealing with severe cases.

Antibiotic Baths: and dips are not generally as effective as antibiotics given by injection. Baths and dips may however, be useful for surface infections.

Dimilin: Effective treatment against anchor worm (Lernaea) and possibly fish lice (Argulus)

Medicated Food: is a useful option for treating bacterial disease, but it does have some disadvantages.

Topical Treatments: can be very effective against bacterial diseases - especially when used in conjunction with antibiotics

Chloramine-T: Can be tricky to use - but is effective against some parasites - it can also assist in gill problems.

Copper: a treatment for marine parasites.

Malachite Green and Formalin: an all-round anti-parasite treatment, especially good for white spot.

Organophosphates: the "bad boy" of treatments - but the best treatment for the "difficult" parasites such as Argulus.

Potassium Permanganate: an effective treatment against many parasites, especially Trichodina. Can be useful for bacterial problems and disinfecting nets etc.

Quaternary Ammonium Compounds: (QACs): useful for gill disease - as it acts as a mild disinfectant with detergent action to remove debris.

Salt: The fish keeper's standby for parasites, gill congestion and osmoregulation problems. It's safe, and cheap

Using "medications"

- If possible, make a definite diagnosis rather than a presumption or guess. This enables specific targeting of both the problem and cause. Follow the diagnosis guide
- Check, double check and get someone else to check if you are not 100% sure of the volume of the pond or tank.
- If there are signs of distress remove the "patients" or rapidly dilute the treated water
- It is always advisable to carry out a follow-up examination to ensure the effectiveness of the medication and if necessary switch to another.

Medicated/antibiotic Treated Food to Treat Bacterial Disease

When a bacterial disease has been diagnosed, one treatment option is to use antibiotics. There are three main routes for antibiotic usage; baths, injections and medicated food. Using medicated food is an option that minimises handling and therefore stress – although affected fish may still need topical treatment for moderate to severe lesions.

One disadvantage of using medicated food is that it is not possible to ensure that individual fish receive the right dose of medication; indeed sick fish may not feed at all. The other problem with this method is the limited choice of antibiotics in ready prepared food – which for hobbyists in the UK is restricted to oxolinic acid. It is possible to "top dress" ordinary food with other antibiotics but this brings with it the problems of palatability and the risk of antibiotics being washed off the food before it is eaten.

Ready-prepared Antibiotic Foods

As already mentioned, in the UK oxolinic acid is the only antibiotic treated food available to hobbyists. Unfortunately, because of previous overuse and misuse, there is a very high level of bacterial resistance to oxolinic acid. Other antibiotics are available in commercial fish foods used to treat food fish such as salmon and trout, but my understanding is that these are not available to hobbyists. All ready-prepared medicated food requires a veterinary prescription – which will involve a visit to your local vet.

- For further details phone: King British Ltd. Mr. Holmes 01274 576241
- For more details on commercial preparations phone: Vetrapharm Ltd. 01425 656081

Commercially Available Medicated Food in America

Top Dressing Ordinary Food: The other option is to apply antibiotic directly to ordinary pelleted food. This increases the options of which antibiotic can be used. Again, the antibiotic will have to be obtained via your local veterinary surgery. The two major problems (once you have calculated the dose) are getting the antibiotic to bind to the food and not float off when the pellets are put in the water, and palatability – that is making sure they still taste nice! Tests carried out at the Fish Disease Laboratory in Weymouth showed either vegetable or fish oils to be the best binding agents and to have the least affect on palatability.

Preparing the Food

Using table 1 calculated the total amount of antibiotic required. First calculate the total weight of the fish to be treated. As a rough guide, a 4" fish will weigh 100 grams. 12" fish will weigh approx. 1kg. 18" fish will weigh approx. 2kg.

Determine the total amount of antibiotic required per day. For example if the total weight of fish was 30kg and you were using oxytetracycline at a dosage of 75mg/ kg body-weight, you would need 30 x 75mg = 2.25 grams of active ingredient per day.

On a 14 day course the total amount of active oxytetracycline required would be 14 x 2.25 grams = 31.5 grams active ingredient (you should be aware that not all antibiotics are 100% pure).

Calculate the amount of food the fish will eat over a 14-day period. A good estimate is 1% total body-weight per day. In our example, 30kg of fish will eat approx. 30kg x 1% x 14 = 4.2 kg food. Use a small amount of oil; say 2-3 grams per 100grams of food. Warm the oil to about 40°C and quickly stir in the total antibiotic for the 14-day period. Add the oil and drug slurry to all the pellets and stir to spread the oil slurry evenly over all the pellets. Cool and divide into 14 equal portions. Store in a sealed container in a cool room or refrigerator. It is sometimes helpful to starve the fish for 24-hours before starting the treatment

Ammonia Poisoning

Disease Type: Environmental

Cause: Unionized Ammonia (NH3)

Description: Ammonia poisoning is one of the biggest killers of aquarium fish. It occurs most often when a tank is newly set up. However, it can also occur in an established tank when too many new fish have been added at one time, when the filter fails due to power or mechanical failure, or if bacterial colonies die off due to the use of medications or sudden change in water conditions.

Symptoms:

- Fish gasp for breath at the water surface
- Purple or red gills
- Fish is lethargic
- Loss of appetite
- Fish lays at the bottom of the tank
- Red streaking on the fins or body

Ammonia poisoning can happen suddenly, or over a period of days. Initially the fish may be seen gasping at the surface for air. The gills will begin to turn red or lilac in color, and may appear to be bleeding. The fish will being to lose its appetite and become increasingly lethargic. In some cases fish may be observed laying at the bottom of the tank with clamped fins. As the damage from the ammonia poisoning continues, the tissues will be damaged as evidenced by red streaks or bloody patches that appear on the body and fins. Internal damage is occurring to the brain, organs, and central nervous system. The fish begins to hemorrhage internally and externally, and eventually dies.

Treatment:

- Lower pH below 7.0
- 25 - 50% water change
- Use chemical to neutralize ammonia
- Discontinue or reduce feeding

If the ammonia level rises above 1 ppm as measured by a standard test kit, begin treatment immediately. Lowering the pH of the water will provide immediate relief, as will a 50% water change (be sure to use water that is the same temperature as the aquarium). Several water changes within a short period of time may be required to drop the ammonia to below 1 ppm. If the fish are in severe distress, the use of a chemical to neutralize the ammonia is recommended. Feedings should be restricted so that additional waste is reduced. In cases of very high ammonia levels, feedings should be discontinued for several days. No new fish should be added until the tank until the ammonia and nitrite levels have fallen to zero.

Because ammonia toxicity is linked to the pH, testing of both ammonia and pH levels are critical. Ammonia becomes increasingly toxic as the pH rises above 7.0. Because there are so many variables, there is no magic number to watch for. However, there are general guidelines to follow.

At a level of level of 1 ppm or 1 mg/l, fish are under stress, even if they don't appear in acute distress. Levels even lower than that can be fatal if the fish are exposed continuously for several days. For that reason it is critical to continue daily testing and treatment until the ammonia drops to zero. When ammonia is elevated for a long period, it is not unusual to lose fish even after the ammonia levels start to drop.

Prevention:

- Stock new tanks slowly
- Feed sparingly and remove uneaten food
- Change water regularly
- Test water regularly to catch problems early

The key to avoiding fish death from ammonia poisoning is to avoid ammonia spikes in the first place. When starting a new tank, add only a couple of fish initially and do not add more until the tank is completely cycled. Even in an well established tank, only add a couple of new fish at a time and avoid overstocking.

Feed fish small quantities of foods, and remove any food not consumed in five minutes. Clean the tank weekly, taking care to remove an dead plants or other debris. Perform a partial water change at least every other week, more often in small heavily stocked tanks. Test the water for ammonia at least twice a month to detect problems before they become serious. Anytime a fish appears to be ill, test for ammonia to rule out ammonia poisoning. If the filter stops, test for ammonia twenty-four hours later to ensure that the bacterial colonies that eliminate wastes were not affected.

The Nitrogen Cycle

How an Aquarium 'cycles'

By Shirlie Sharpe, About.com Guide

- ammonia poisoning
- nitrites
- cloudy water
- water changes
- aquarium bacteria

Freshwater Aquariums Ads

- Fish Aquarium Supplies
- Marine Fish Tank
- Freshwater Aquarium
- Aquarium Tanks
- Cloudy Water in Fish Tank

Call it cycling, nitrification, biological cycle, startup cycle, break-in cycle, or the nitrogen cycle. No matter what name you use, every newly set up aquarium goes through a process of establishing beneficial bacterial colonies. Older aquariums also go through periods during which the bacterial colonies fluctuate. Failure to understand this process is the largest contributing factor to the loss of fish. Learning what it is, and how to deal with critical periods during the nitrogen cycle, will greatly increase your chances of successful fish keeping.

The Waste Problem

Unlike nature, an aquarium is a closed environment. All the wastes excreted from the fish, uneaten food, and decaying plants stay inside the tank. If nothing eliminated those wastes, your beautiful aquarium would turn into a cesspool in no time at all.

Actually, for a short period of time, a new aquarium does become a toxic cesspool. The water may look clear, but don't be fooled. It's loaded with toxins. Sounds awful, doesn't it? Fortunately bacteria that are capable of converting wastes to safer by-products begin growing in the tank as soon as fish are added. Unfortunately there aren't enough bacteria to eliminate all the toxins immediately, so for a period of several weeks to a month or more, your fish are at risk.

However, you need not lose them. Armed with an understanding of how the nitrogen cycle works and knowing the proper steps to take, you can sail through the break-in cycle with very few problems.

Stages of the Nitrogen Cycle

There are three stages of the nitrogen cycle, each of which presents different challenges.

Initial Stage: The cycle begins when fish are introduced to the aquarium. Their feces, urine, as well as any uneaten food, are quickly broken down into either ionized or unionized ammonia. The ionized form, Ammonium (NH4), is present if the pH is below 7, and is not toxic to fish. The unionized form, Ammonia (NH3), is is present if the pH is 7 or above, and is highly toxic to fish. Any amount of unionized Ammonia (NH3) is dangerous, however once the levels reach 2 ppm, the fish are in grave danger. Ammonia usually begins rising by the third day after introducing fish.

Second Stage: During this stage Nitrosomonas bacteria oxidize the ammonia, thus eliminating it. However, the by-product of ammonia oxidation is nitrite, which is also highly toxic to fish. Nitrites levels as low as low as 1 mg/l can be lethal to some fish. Nitrite usually begins rising by the end of the first week after introducing fish.

Third Stage: In the last stage of the cycle, Nitrobacter bacteria convert the nitrites into nitrates. Nitrates are not highly toxic to fish in low to moderate levels. Routine partial water changes will keep the nitrate levels within the safe range. Established tanks should be tested for nitrates every few months to ensure that levels are not becoming extremely high.

Now that you know what is happening, what should you do? Simple steps such as testing and changing the water will help you manage the nitrogen cycle without losing your fish. For details about what to do next, continue to Page 2.

Managing the Nitrogen Cycle

Freshwater Aquariums Ads

- Aquarium
- Aquarium PH Test
- Fish Tanks
- Freshwater Fish
- Fish Aquarium Supplies

What to Do

The key for success is testing the water for ammonia and nitrites, and taking action quickly when problems occur. To aid in tracking the status of your aquarium, links to charts for logging your tests can be found under the charts section of this page. Each chart shows the danger zones and offers steps to reduce toxins before they result in loss of your fish.

Test for Ammonia: Begin testing on day three after adding the fish, and continue every day until the ammonia begins to drop. After it begins to fall, continue testing every other day until the ammonia reaches zero. Using the chart provided, plot the ammonia levels. Should ammonia reach the danger zone, take steps as shown on the chart. If at any time fish show signs of distress, such as rapid breathing (gilling), clamped fins, erratic swimming, or hanging at the surface for air, take immediate action to lower the ammonia level. Chemicals such as Ammo-Lock will quickly neutralize toxic ammonia.

Test for Nitrites: Begin testing one week after adding the fish. Continue testing every second or third day, until it reaches zero. Using the chart provided, plot the nitrite levels and take steps as shown on the chart if nitrite reaches the danger zone. If at any time fish show signs of distress, such as rapid breathing or hanging near the surface seemingly gasping for air, test for nitrite. If levels are elevated perform an immediate 25-50% water change and test daily until levels drop.

What Not to Do

- Don't add more fish - wait until the cycle is completed.

- Don't change the filter media - the beneficial bacteria are growing there. Don't disturb them until they have become well established.
- Don't overfeed the fish - when in doubt underfeed your fish. Remember that anything going into the tank will produce wastes one way or another.
- Don't try to alter the pH - the beneficial bacteria can be affected by changes in pH. Unless there is a serious problem with the pH, leave it alone during the startup cycle process.

Columnaris - Flexibacter Columnaris

Disease Type: Bacterial (gram negative rods)

Organism: Flexibacter columnaris

Names: Columnaris, Cotton-Wool, Cotton-Mouth, Flexibacter, Mouth Fungus

Description

Often mistaken for a fungal infection because of its mold-like lesions, Columnaris is a common bacterial infection in cultured fish, particularly livebearing fish and catfish. Its name is derived from columnar shaped bacteria, which are present in virtually all aquarium environments.

The bacteria are most likely to infect fish that have been stressed by conditions such as poor water quality, inadequate diet, or stress from handling and shipping. Columnaris can enter the fish through the gills, mouth, or via small wounds on the skin. The disease is highly contagious and may be spread through contaminated nets, specimen containers, and even food. For this reason it is important to use sterile techniques to avoid contaminating other tanks. Prophylactically treating other tanks is wise, especially if they share a common filtration system.

Columnaris can be external or internal and may follow a chronic or acute course. Lesions in chronic cases progress slowly, taking many days before culminating in fish death. In acute cases the lesions spread quickly, often wiping out entire populations of fish within hours. High water temperatures accelerate the progression of the disease; however lowering the water temp will not affect the outcome of the disease.

Symptoms

- White spots on mouth, edges of scales, and fins
- Cottony growth that eats away at the mouth
- Fins disintegrate beginning at the edges
- 'Saddleback' lesion near the dorsal fin
- Fungus often invades the affected skin
- Rapid gilling in cases where gills are infected

Most Columnaris infections are external, and present first as white or grayish white spots on the head, and around the fins or gills. The lesions may first be seen only as a paler area that lacks the normal shiny appearance. As the lesion progresses it may become yellowish or brownish in color and the area around it may be tinged red.

Lesions on the back often extend down the sides, giving the appearance of a saddle. On the mouth the lesions may look moldy or cottony, and the mouth will become eaten away. Fins will erode and have a frayed appearance as the infection progresses. The gills filaments will disintegrate as the bacteria invade them, and the fish will begin breathing rapidly due to lack of oxygen. Less commonly, the infection will be internal, and display no external symptoms. In these cases, only a necropsy and cultures will point to the true cause of death.

Treatment

- Change water
- Vacuum gravel
- Add aquarium salt
- Treat with copper sulfate or antibiotic
- Discontinue carbon filtration during treatment

External infections should be treated with antibiotics or chemicals in the water. Copper sulfate, Acriflavine, Furan, and Terramycin may all be used externally to treat Columnaris. Terramycin has proven to be quite effective both as a bath, and when used to treat foods for internal infections. Salt may be added to the water to enhance gill function. Livebearers in particular will benefit from the addition of salt, however use caution when treating catfish, as many are sensitive to salt. When in doubt error on the side of caution when using salt.

Prevention

- Quarantine new fish for two weeks
- Maintain high water quality
- Provide fish with a nutritionally balanced diet
- Medicate fish prophylactically before moving them
- Disinfect nets and other equipment before using

Because the bacteria thrive on organic wastes , it can be controlled by regular water changes and vacuuming of the gravel. Proper diet and maintaining good water quality in general will keep the fish from being stressed and therefore susceptible to infection. To avoid spreading the bacterium, nets, specimen containers, and other aquarium equipment should be disinfected before each use. Small quantities of aquarium salt can be used to prevent disease in livebearer aquariums. When fish are being shipped or moved, they may be treated prophylactically with antibiotics or by feeding them medicated food.

Velvet - Oodinium Pilularis

Freshwater Aquariums Ads

- Sick Fish Treatment
- Fish Aquarium Supplies
- Marine Aquarium
- Koi Disease
- Fish

Overview

- *Disease Type:* Parasitic skin flagellate
- *Organism:* Oödinium pilularis
- *Names:* Rust, Gold Dust Disease, Oödinium, Velvet

Description

The name sounds nice, but don't be fooled. Velvet is one of the more common diseases in aquarium fish, and can strike down every inhabitant in the tank before the hapless owner realizes what he or she is dealing with. Also known as Rust or Gold Dust disease, it is caused by one of several species of a tiny parasite known as Oödinium.

Oödinium is a dinoflagellate, a creature classified by some as a protozoan and by others as algae because it contains Chlorophyll. Oödinium doesn't care how it's classified; it's an equal opportunity parasite that strikes fresh and saltwater fish.

In freshwater fish Velvet is caused by either Oödinium pilularis or Oödinium limneticum. In marine fish Oödinium ocellatum causes the dreaded Coral Fish Disease. All three species have symptoms and lifecycles similar to the well known parasite, Ich.

Oödinium finds a fish and adheres to it using flagellum, then forms rod pseudopodia which penetrate the skin and soft tissues of the gills. The pseudopods destroy the cells and feed on the nutrients inside. After feeding and maturing, the parasite drops off the fish and divides into dozens of cells that are released into the water to seek hosts. They must find a host within 24 hours, or die.

Oödinium produces white pustules on the fish that are much finer than the spots seen in Ich. In fact they are so fine they are often not seen before the fish perishes. Like Ich, Oödinium is present in most commercial tanks, but only becomes a problem when the fish are stressed by poor quality water, changes in the water temperature, or being transported.

Symptoms

- Scratchs against hard objects
- Fish is lethargic
- Loss of appetite and weight loss
- Rapid, labored breathing
- Fins clamped against body
- Fine yellow or rusty colored film on skin
- In advanced stages skin peeks off

Initially the fish rub against hard objects trying to dislodge the parasites. As the disease progresses the fish becomes lethargic, fins are held close to the body, appetite is reduced and the fish loses weight. A key symptom is difficult breathing, resulting in rapid gilling. Perhaps the most telltale symptom is the appearance of a velvety film on the skin that resembles gold or rust colored dust. The film may be difficult to see, but can be more easily detected by directing a beam of a flashlight on the fish in a darkened room. The parasite is most often seen on the fins and gills. Velvet attacks all fish and will even affect fry that are only a few days old. Anabantoids, Danios, Goldfish, Zebras, and Killifish are particularly susceptible to velvet disease.

Treatment

- Raise water temperature
- Dim lights for several days
- Add aquarium salt
- Treat with copper sulphate for ten days
- Discontinue carbon filtration during treatment

Because Velvet is highly contagious and usually far advanced before being diagnosed, it is important to take steps to treat it as soon as possible. Treatment is targeted at the free-swimming stage of the parasite. Copper sulphate is the treatment of choice. It should be used according to the manufacturer's instructions for a full ten days to ensure that the parasite is completely eradicated. Atabrine (Quinacrine hydrochloride) is another medication that can be used to treat Velvet.

Because Oödinium is dependant on light, dimming the aquarium lights aids in eliminating the infestation. Increasing the water temperature to 82°F will speed the process, and adding salt to the water will ease the labored breathing caused by destruction of gill tissue. As with any treatment, activated carbon should be removed from the filter, as it will remove the drugs from the water.

Prevention

- Quarantine new fish for two weeks
- Maintain high water quality
- Provide fish with a nutritionally balanced diet

Velvet usually only arises when poor aquarium conditions prevail and is highly infectious. Quarantine of new fish for two weeks will greatly reduce the likelihood of contaminating a healthy established aquarium. Any fish that appear to be ill should immediately be removed and kept in a hospital tank to avoid the spread of the parasite.

Top Ich and Other External Fish Parasite Treatments

Medications Effective Against Parasitic Infestions

Parasitic infestations in a saltwater aquarium are among the most feared and difficult to treat. There are a number of medications on the market designed to prevent and/or eliminate parasites. Some can be used with invertebrates, some can not.

Chem-Marin Stop Parasites

Natural ingredients remove all types of parasites, including saltwater Ich. Completely safe for all fish, invertebrates, and corals and does not harm the fish's liver as does copper. This organic solution is 99% effective in fresh and saltwater aquariums. Stop Parasites works by first enhancing the slime coat on fish, which will remove all parasites. Then it uses another additive, which the parasites attach themselves to as a false host, and the parasites are then naturally removed through your protein skimmer or mechanical filtration. Another agent in the product works to speed up appetite and build the immune system.

CopperSafe

Manufactured by Mardel Laboratories.

Stabilized copper that effectively treats external ich, velvet, flukes, and other parasites. Maintains a therapeutic level of copper that is safe for fish, but is effective against parasites.

Not Reef safe

For use in freshwater or saltwater fish aquariums only.

One dose treats for one month.

Manufactured by Seachem.

This buffered, active copper is claimed to be safer than copper sulfate or chelates for eradicating ectoparasites.

4X concentration gap between the minimal therapeutic dose and the toxic dose.

Fully ionic and effective at low concentrations.

Safe for filter bed. Can be removed with chemical filtration.

Not for Use with Invertebrates

Manufactured by Kordon.

100% organic (multiple natural herbals containing Naphthoquinon).

Active ingredients treat diseases caused by Ich, fungus, protozoans, and dinoflagellates.

Does not affect beneficial nitrifying bacteria.

Can be used with all Kordon Water Conditioners such as NovAqua, AmQuel, and PolyAqua, all Kordon Water Clarifiers (TransClear and Sea Clear) and all AquaTru Water Quality Test Kits.

Manufactured by Hikari

Treatment for Ich, cryptocaryon, velvet, and Ich related conditions.

Malachite green formula provides proven healing, and slime coat replacer speeds recovery and reduces further infection.

Use with Caution in Aquariums with Invertebrates

Ingredients: Water, formaldehyde (<5%), methanol (<2%), malachite green chloride (<0.1%).

This treatment, Manufactured by Aquarium Systems, uses a new method to cure parasites. Broad-spectrum treatment effective against marine ich and oodinium.

Non-antibiotic tablets rely on powerful oxidizing action to attack and destroy a wide range of disease-causing microorganisms.

Treats infectious ailments such as ich, oodinium, and more.

Place tablets in floating dispenser to allow medication to dissolve and distribute evenly throughout the aquarium.

Manufactured by Kordon

Treats saltwater external parasites such as Cryptocaryon and Amyloodinium. Also treats fungal infections in fishes.

Contains formaldehyde and a zinc-free chloride salt of malachite green.

Not Recommended for Use in Aquariums Containing Invertebrates

Reef Safe Kick-Ich is a water treatment for the control of ich in marine and freshwater aquaria. It has been scientifically formulated to eliminate the free swimming, infectious stage of the ich life cycle.

Ruby Reef Rally

Reef Safe Rally is a copper free water treatment. It can be used in any tropical or marine aquarium with harmful bacteria, flukes, or oodinium Safe for plants and invertebrates.

Manufactured by Aquarium Systems

Standardized copper sulfate medication.

Controls saltwater velvet and Ich.

Can be monitored with a copper test to ensure safe and effective treatment.

For Fish Only Tanks (not for use in tanks with invertebrates).

Pharmaceutical grade active ingredients for maximum efficacy.

Quick-dissolving powder for faster treatment of aquarium ich.

Human warning: Carcinogenesis: Nitrofurazone, an active ingredient in Super Ick Cure, has been shown to produce mammary tumors on rats and ovarian tumors on mice.

Active ingredients: 3.6 mg Malachite Green and 60 mg Nitrofurazone per packet.

Most Common Mistakes Made by Saltwater Aquarium Keepers

Overfeeding Fish and Invertebrtes

Uneaten food just lays on the bottom of the tank, creating nitrates and overloading the biological filter.

Not fully understanding the nutritional requirements of their fish, the tendency of many people is to "throw food" at fish in order to fulfill their requirements. If the fish are not accepting the food offered, many aquarists will "throw even more" at the fish, thinking that the fish just isn't seeing the food. Feed once, twice per day, or once every 2 or 3 days? How Often Should I Feed My Fish? helps you understand a fish's requirements.

Know what is in the food you are feeding by comparing the nutrients in commercial foods, purchase only high quality foods and feed only what your fish will consume in 2-3 minutes per feeding.

Moving Too Fast

"Patience" is a requirement with just about anything that you do with a saltwater aquarium. Far too many people report problems after they have put a tank together, because they are just moving too fast! Far too often we have read aquarists comments like, "I need test kits? What for, and what kind?" Of course this is after they have had a tank for some time. A high percentage of people do not take the time to read and study up on the hobby before getting started.

Overloading the System

A problem that goes hand-in-hand with moving too fast is craming too much livestock and/or live rock into the aquarium all at once, especially in a tank that is not fully cycled, or has just completed the cycling process. Even in a well established system, placing too many new additions into the tank to quickly can cause new tank syndrome. Slow down! Saltwater aquarium keeping is not a timed event, so take it easy, and work on your patience skills.

Inadequate Filtration and Water Circulation

Having sufficient biological filtration is a primary key to success in keeping a saltwater aquarium. There are a number of filtration methods to choose from, but not making the right filter selection for the bio-load planned for your tank can lead to a wide variety of problems. Whether it be biological, mechanical, or chemical, it's better to have more, rather than too little filtration. This same concept applies to circulation of the water in the aquarium as well. The lack of good water flow throughout the system can lead to problems with low DO (dissolved oxygen), the build up of slime or other types of nuisance algae, prevention of stationary animals receiving food, and more. The solution here? Add a powerhead or two, or a surge device.

Misdiagnosing Diseases

When it comes to diagnosing diseases, saltwater ich is the biggest problem. It is easy to confuse Oodinium (Amyloodinium ocellatum - a.k.a. Marine Velvet or Coral Fish Disease) with White Spot Disease (Cryptocaryon irritans). They are similar but two quite different types of saltwater ich, and each responds to different types of treatment. It is important to properly diagnose and treat these parasites, as well as other diseases.

Overmedicating

Way too often one or more remedies are just thrown at a sick or ailing fish without knowing what the problem is. Medications should only be used when necessary, and whenever possible in a quarantine tank. The most important factor with medications is to use one that is formulated to "target" the specific disease or diseases you are dealing with.

Purchasing Animals Without Knowing Anything About Them

It never ceases to amaze us how often people select new additions for their aquarium without knowing what the animals are, how to care for and feed them. Before purchasing anything, take the time to obtain information about it first. You shouldn't buy on impulse because you like the pretty colors a fish has, how cute or stunning it looks, or for any other "touchy-feely" reason, or if a sales person can't provide you with critical information you need to know about a particular animal.

Livestock Incompatibility

Statements like *my wrasse ate my hermit crab, my tangs just won't get along,* and similar ones are all too frequently heard. Purchasing livestock without knowing whether or not they will peacefully reside with other tankmates can lead to dead or injured animals, as well as stress related diseases. Use common sense and learn about the compatibility of animals you are considering for your aquarium, before putting them together!

Purchasing Animals in Poor Health

One of the easiest things to do when selecting a critter is to determine whether or not it is healthy. In a simple phrase, most sick fish don't eat. Before purchasing a fish or other animal, it is best to have a sale's person in a store show you that it is in fact eating. On your part, learn how to recognize the symptoms or outward signs of common illnesses so you know what to look for when inspecting livestock to buy.

Using a Poor Quality Fresh Water Source

Although many aquarists do so, choosing to use water straight from the tap or unpurified water of another source to make up saltwater solutions and to top off a tank can lead to many water quality issues in aquariums. Using a water purification filter, buying clean natural sea water, or prefiltered RO/DI water from a reliable supplier is an investment that will pay for itself in the long run.

Lack of Proper Tank Maintenance

Well-maintained saltwater systems seldom experience high nitrate, bacterial outbreaks, or other water quality issues. To avoid the usual pitfalls with problems in this area of aquaria keeping, set up and follow a regular maintenance routine.

Ichthyophthirius Multifilis – Ich

Freshwater Aquariums Ads

- Sick Fish Treatment
- Fish
- Fish Aquarium Supplies
- Marine Fish Tank
- Koi Disease

Disease Profile

Type: Parasitic

Organism: Ichthyophthirius multifilis

Names: Ich, Ick, White Spot

Description

The name Ichthyophthirius multifilis translates to "fish louse with many children", a title that fits well, as each parasite may produce over a thousand offspring. Although the disease is the equivalent of a skin infection, ich can easily be fatal to a fish stressed by poor diet or habitat.

Symptoms

- Small white spots resembling sand
- Fish scratch against rocks and gravel
- In advanced stages fish become lethargic
- Redness or bloody streaks in advanced stages

Infected fish are covered to various degrees with small white spots. Severe infestations are easy to spot, but small occurrences often go unnoticed. However, ich won't remain unnoticed for long. Like a bad penny, it will be back with a vengeance.

The adult parasite burrows into the skin of its victim, feeding on blood and dead epithelial cells. The irritation caused by the burrowing parasite causes the skin of the fish to swell and produce white cysts seen as a small spots. The fish feels as if it's been bitten by a mosquito. It's not unusual to see infected fish scratching against rocks and gravel in an effort to get relief.

After several days of feasting, the engorged parasite develops into a trophozoite, burrows out of the fish and sinks bottom of the tank.

Secreting a soft jellylike substance, it forms a protective membrane inside of which it divides into hundreds of baby parasites, known as tomites. The hungry tomites soon leave their home in search of a fresh fish to dine upon.

It is during the free-swimming stage, which lasts a mere three days, that the parasite is vulnerable to medication. Once it has burrowed into a new host fish it is safely protected from chemicals in the water.

Treatment

- Raise water temperature
- Medicate for 10-14 days
- Reduce medication when treating scaleless fish
- Discontinue carbon filtration during treatment
- Perform water changes between treatments

The entire cycle takes about two weeks from start to finish. Higher temps will shorten the cycle, while low temps lengthen it. Therefore, raising the water temp shortens the time it takes for the parasite to reach the stage in which it is susceptible to medication.

Treatments must be given for a long enough period to assure that all parasites are gone. Watch carefully for other infections, as secondary infections often occur where the skin has been damaged by the parasite. Although nothing kills the parasite once it has checked into it's fish "hotel", several chemicals kill ich once it has left the fish. Malachite green, methylene blue, quinine hydrochloride, and mepracrine hydrochloride are all effective, and are available under several brand names.

Dose based on the package instructions, however cut dosages in half when treating scaleless catfish and tetras. Regardless of the medication used, treatment should be given continuously for 10-14 days to ensure all parasites are killed.

Between treatments a partial water change is recommended. Keep water temperatures higher than usual to speed up the life cycle of the parasite. Discontinue carbon filtration during treatment, as it will remove the chemicals.

Prevention

- Quarantine new fish for two weeks
- Treat plants before adding to tank
- Maintain high water quality
- Provide fish with a nutritionally balanced diet

The best way to avoid ich is to quarantine all new fish in a separate tank for two weeks before moving them to the regular tank. When quarantine is not possible, a prophylactic treatment may be used. Either methylene blue or malachite green given when new fish are introduced and again four days later will help reduce incidence of infection. New plants should also be treated, as they can carry ich cysts.

Maintaining high water quality, avoiding temperature fluctuations,and providing a robust diet is the best preventative for ich and other diseases.

4

Saltwater Ich/White Spot Disease

Diagnose and Treat Cryptocaryon

Saltwater Aquariums Ads

- Sick Fish Treatment
- Marine Aquarium
- Marine Fish Tank
- Fish Swimming
- Saltwater Aquarium

What is Cryptocaryon?

Saltwater Ich or White Spot Disease is caused by an infestation of the ciliated protozoan *Cryptocaryon irritans*. Although just like the ich organisms Oodinium (Marine Ich, Velvet or Coral Fish Disease) and Brooklynella (Clownfish Disease) that also become parasitic to fish at one stage in their life cycle, Crypto progresses less rapidly. If detected early and treated promptly upon an outbreak, the chances of recovery are high. However, in a closed aquarium system it can reach overwhelming and disastrous numbers just the same if it is not taken care of.

The Life Cycle of Cryptocaryon irritans

- Free-swimming cells called *tomites* are released from a mature tomont, or encrusted cyst, and go in search of a host fish, typically dying in a day or two if one is not found.
- Upon finding a host the tomites attach to the gills or body and develop into parasitic *trophonts*, at which stage the organisms burrow into the fish and begin feeding on its tissues.

- Once well fed the trophonts stop feeding and encyst, at which stage they become inactive *tomonts*. These dormant cysts can remain trapped in the fish's mucus, be inbedded deep in the tissue, or drop off and fall to the bottom. Over a period of 6 to 10 days the cells inside the cysts reproduce by single-cell division, and become tomites. Once reaching maturity the cysts rupture, each releases hundreds of new free-swimming tomites, and the cycle begins again, but in much larger numbers.

Symptoms

Unlike Oodinium and Brooklynella that typically attack the gills first, which allows these ich diseases to advance into life-threatening levels quickly as they go unnoticed, Cryptocaryon usually appears at the onset as salt-sized white spots visible on the body and fins of a host fish, and when the organisms become parasitic, it is then that they move inwards to the gills. Because crypto is more easily recognized in its beginning stage, this makes it much easier to treat and cure before it gets out of control.

Aside from the appearance of the white spots, fish will scratch against objects in an attempt to dislodge the parasites, and rapid respiration develops as tomonts, mucus, and tissue debris clogs the gills. Fish become listless, refuse to eat, loss of color occurs in patches or blotches as the trophonts destroy the pigment cells, and secondary bacterial infections invade the lesions caused by the trophonts.

Treatment Recommendations

Although copper is very effective on Oodinium, and it works well to eliminate crypto organisms in their free-swimming tomite stage, it is not as effective on the Cryptocaryon trophonts that burrow deeply into the tissues of fish. A combination of freshwater and formalin treatments administered by means of dips, baths, and prolonged treatment over a period of time in a QT is recommended.

How to Treat Ich Diseased Fish with Formalin

Preventing Reinfestation: Reinfection will occur no matter how effectively the fish have been treated if Cryptocaryon is not eradicated from the main aquarium, which can be accomplished by keeping the tank devoid of any fish for at least 4 weeks. For fish-only aquariums hyposalinity can be applied, and to speed up the life cycle of the organisms, elevate the tank reef aquariums, Ruby Reef Kick-Ich and Chem-Marin Stop Parasites are "reef safe" formulated treatments that specifically target the Crytpocaryon organims.

Several days prior to returning fish to the main aquarium, clean all filtering equipment, change any filtering materials, and perform a water change.

Medication Usage Tips:

- Porous materials such as sand, gravel, rocks, and ornaments can absorb medications. To better control the strength and effectiveness of any product you are using, it is best to use a bare QT with only some cut pieces of PVC tubing in the tank to provide shelter for the fish during the treatment period.
- Although many over-the-counter remedies contain the general name Ich or Ick, carefully read the product information to be sure it is designed to specifically target and treat "Cryptocaryon".

Saltwater Aquarium Emergency First Aid Kit

Unforeseen situations with your aquarium can occur at any time, and if you are not prepared to handle a critical problem when (not if) it arises, losses to your aquarium community can be costly. You should be ready at any given time to be able to perform a partial or complete water change, treat a sick fish, or handle any number of other emergencies.

In some situations, immediate action may need to be taken and time to waste is something you may not have. For this reason we feel it is important to keep an emergency first aid kit on hand, here is a list of basic items we feel an aquarists should have readily available.

Quarantine Tank

Every saltwater aquarist should have an easily set up quarantine tank. Is a QT Really Necessary? explains why. In a pinch, How to Create a Fast & Cheap QT offers a number of solutions.

Test Kits

By testing your aquarium water regularly steps can be quickly taken to prevent or head off potential problems. Water tests give you a chemical snapshot of what is going on in your tank water.

- Why test aquarium water?
- Read Test Kit Reviews & Compare Prices
- Test Kit Polling Results

Baking Soda

Sodium bicarbonate (baking soda) can be used as a quick band-aid to raise the pH in a tank. Found in most kitchens, a dose of baking soda will temporarily suffice in an emergency situation until a water change can be made. Just dissolve one teaspoon of baking soda for each 20 gallons of tank water in a cup of water, then slowly pour it into your aquarium. The pH change will not be noticeable immediately, so wait for about an hour and test again before adding more. Repeat until you obtain the desired pH level.

Ammonia Neutralizer

When you notice that your fish suddenly have frayed fins and/ or cloudy eyes the probable cause is excess ammonia in your tank water. Quickly dosing your tank with a good ammonia neutralizer is the first step to limit the damage. Ammonia neutralizers are also useful in new Quarantine Tanks (QT) until a mature biological filter can be established.

Water Conditioners

These products help restore a fish's protective slime coat which is important after an injury has occurred. Many of these products also neutralize ammonia or remove chlorine and chloramines from tap water.

Antibacterials/Antibiotics

The most important thing to remember about medications is that many are not reef safe, and some can weaken or completely kill the biological filter in an aquarium, so it is best to treat with any type of medication in a QT, NOT in the main aquarium. Read and follow the directions for recommended dosages and proper protocol closely. Also check to see if they have a shelf life. Since the medications are going to be kept in your first aid kit for emergencies, you want to make sure their effectiveness does not expire.

Treat and prevent the bacterial infections which can follow injuries, pop eye, frayed fins and secondary parasite related injuries.

Ich Treatments

Treating parasites as soon as they are detected will greatly increase the success of a cure as well as reduce the probability of fatalities. Keep a supply of medications for both Oodinium (Velvet/Rust/Coral Fish Disease) and Cryptocaryon (White Spot Disease). Most aquarists experience both of these parasites at one time or another.

Red Slime Algae Treatment

Red Slime Algae can pop up from time to time. Red Slime Algae can quickly take over an entire aquarium, choking off the substrate, live rocks and corals. Having an effective treatment on hand when it starts will help you get rid of it quickly.

Extra Sea Salts

Having an extra week or 2 supply of sea salts held for emergency water changes or setting up a quick QT is always a good idea.

Extra Aquarium Heater

Maintaining a stable water temperature is important for aquarium occupants. Aquarium heaters can and do malfunction without notice. Keeping a spare heater in your emergency kit will help minimize a potential disaster in the future.

Activated Carbon

Keeping a couple of pounds of activated carbon in your kit will allow you to quickly filter toxins out of your aquarium water if the need arises.

Saltwater Aquarium Antibiotic Products

Top Picks

When a bacterial outbreak occurs in a saltwater aquarium, quick treatment with the appropriate antibiotic can make the difference between an effective cure and a total wipe out of aquarium livestock.

Furan-2

Furan-2 by Aquarium Pharmaceuticals which claims: "Effectively treat a wide variety of gram-positive and gram-negative bacterial diseases including: furunculosis (aeromonas), dropsy, gill disease, fin and tail rot, open red sores, black molly disease, mouth fungus, hemorrhagic septicemia, body slime and eye cloud. Powder dissolves quickly for faster healing. Active ingredients: Nitrofurazone and Furazolidone."

Maracyn Plus

By Mardel which claims: "Revolutionary antibiotic medication for effective treatment of fish. Medicated bio-spheres attach to the fish's skin to form a protective layer that kills infecting bacteria. Controlled, regulated release of antibiotics helps prevent re-infection. Maracyn Plus contains two broad-spectrum antibiotics that are effective against fin and tail rot, popeye, dropsy, ulcers, and mouth and body fungus."

Maracyn-Two

Maracyn-Two by Mardel is a broad-spectrum antibiotic for internal or external gram-negative bacterial infections. Effective treatment for fin and tail rot, popeye, gill disease, dropsy, bleeding or red streaks, secondary and internal infections.

Maracyn

Maracyn, by Mardel is used in the treatment of body fungus, fin and tail rot, popeye, gill disease, secondary infections and other gram-positive bacterial infections and fungal diseases.

MelaFix

By Aquarium Pharmaceuticals which claims: "Contains the natural botanical extract from the Tea Tree (Melaleuca), an excellent alternative to resistant strains of bacteria that are unaffected by traditional medications. Treats bacterial infections such as red ulcers, fin and tail rot, cloudy eyes, mouth fungus, and others in as little as 4 days. Also heals open wounds, ulcers, and damaged fins. Doesn't affect pH; safe for invertebrates."

PimaFix

Pimafix, by Aquarium Pharmaceuticals which claims it "treats cotton-like fungal infections, and both internal and external bacterial infections. PimaFix harnesses the unique antifungal and antibacterial properties of the West Indian Bay Tree (Pimenta racemosa) for a safe and natural remedy; prevents the development of resistant strains of disease-causing organisms. Will not discolor water, affect biological filter, or pH during treatment. Will not harm aquatic plants. For extreme cases, PimaFix can safely be used with MelaFix to provide the added benefit of quick tissue regeneration and wound healing."

Powder Blue Tang Profile

Saltwater Aquariums Ads

- Fish
- Fish Aquarium Supplies
- Marine Fish Tank
- Tropical Fish Pictures
- Salt Water Fish

Scientific Name: *Acanthurus leucosternon* (Bennett, 1832).

Other Common Names: Powder Blue Surgeonfish.

Distribution: Indian Ocean.

Average Size: 9.1 inches (23 cm).

Characteristics and Compatibility: A very popular fish with aquarists, but one that is not easy to care for, the Powder Blue Tang requires much attention. It is highly susceptible to contracting ich, and can have problems with HLLE. It can be aggressive towards other surgeonfishes, particularly those of the same shape and color. It is best housed as the only tang present in an aquarium, with the possible exception of a larger system. Females are considerably larger than males. Can often be delicate to acclimate.

Diet and Feeding

Primarily a herbivore that will graze on filamentous micro and some types of smaller fleshy macroalgae. Should be fed a diet of frozen and dried fares suitable for herbivores that contain marine algae and Spirulina (blue-green algae). Zucchini, broccoli, leaf lettuce, and nori (dried seaweed) can be offered to supplement its diet. Should be fed at least 3 times a day.

Habitat

A somewhat shy fish, the Powder Blue Tang should be provided with ample room to move around and lots of places to hide. Does best in a well established aquarium with an ample growth of algae present to graze on at its leisure.

Suggested Minimum Tank Size: 100 gallons (379 L).

Reef Tank Suitability: Considered safe, but an individual may pick at large polyped stony corals if not well fed.

Guide Notes: It sounds silly, but this a fish that seems to break out with ich if you even look at it the wrong way. Can be a picky eater, and may not readily accept foods offered. For this reason it is best to ask to see this fish eating, before you buy one.

Nitrite Poisoning

Freshwater Aquariums Ads

- Fish Aquarium Supplies
- Fish Tanks
- Freshwater Fish
- Sick Fish Treatment
- Salt Water Fish

Disease Type

Environmental:

Cause: Nitrite

Names: Brown Blood Disease, Nitrite Poisoning

Description

Nitrite poisoning follows closely on the heels of ammonia as a major killer of aquarium fish. Just when you think you are home free after losing half your fish to ammonia poisoning, the nitrites rise and put your fish at risk again. Anytime ammonia levels are elevated, elevated nitrites will soon follow. To avoid nitrite poisoning, test when setting up a new tank, when adding new fish to established an tank, when the filter fails due to power or mechanical failure, and when medicating sick fish.

Symptoms

- Fish gasp for breath at the water surface
- Fish hang near water outlets
- Fish is listless
- Tan or brown gills
- Rapid gill movement

Also known as 'brown blood disease' because the blood turns brown from a increase of methemoglobin. However, methemoglobin causes a more serious problem than changing the color of the blood. It renders the blood unable to carry oxygen, and the fish can literally suffocate even though there is ample oxygen present in the water.

Different species of fish tolerate differing levels of nitrite. Some fish may simply be listless, while others may die suddenly with no obvious signs of illness. Common symptoms include gasping at the surface of the water, hanging near water outlets, rapid gill movement, and a change in gill color from tan to dark brown.

Fish that are exposed to even low levels of nitrite for long periods of time suffer damage to their immune system and are prone to secondary diseases, such as ich, fin rot, and bacterial infections. As methemoglobin levels increase damage occurs to the liver, gills and blood cells. If untreated, affected fish eventually die from lack of oxygen, and/or secondary diseases.

Treatment

- Large water change
- Add salt, preferably chlorine salt
- Reduce feeding
- Increase aeration

The addition of one half ounce of salt per gallon of water will prevent methemoglobin from building up. Chlorine salt is preferable, however any aquarium salt is better than no salt at all. Aeration should be increased to provide ample oxygen saturation in the water. Feedings should be reduced and no new fish should be added until the tank until the ammonia and nitrite levels have fallen to zero.

Nitrite is letal at much lower levels than ammonia. Therefore it is critical to continue daily testing and treatment until the nitrite falls to zero.

Prevention

- Stock new tanks slowly
- Feed sparingly and remove uneaten food
- Change water regularly
- Test water regularly to catch problems early

The key to elminating fish death is to avoid extreme spikes and prolonged elevation of nitrites. When starting a new tank, add only a couple of fish initially and do not add more until the tank is completely cycled. In an established tank, only add a couple of new fish at a time and avoid overstocking. Feed fish small quantities of foods, and remove any food not consumed in five minutes. Clean the tank weekly, taking care to remove an dead plants or other debris. Perform a partial water change at least every other week, more often in small heavily stocked tanks. Always test the water for nitrite after an ammonia spike has occured as there will be a nitrite increase later.

How to Dispose of Deceased Fish

It may sound a bit ridiculous, but you would be surprised at the number of people who write to us asking what to do with a fish that has died. For a beginner in the hobby, it isn't really all that stupid of a question. When a fish store sells the newbie a $200.00 Queen Angelfish to cycle the newly purchased tank he or she is just starting with, we are pretty sure they didn't utter the words, "and when it dies, you should..."

If a fish in your tank dies, to be sure you want to determine the cause of death. Critters don't normally die for no reason at all, with the exception of maybe old age. Some of the causes that can contribute to illness include bacterial diseases, parasites, and poor water quality. It is important before introducing new occupants into the aquarium that the cause of an illness has been eliminated. Otherwise, the underlying cause may be carried over to new introductions, and in all likelihood they will share the same fate as the first.

If a sick fish isn't quite dead and you have exhausted all possible treatments to save it with no results, you may want to consider euthanasia to put it out of its misery. Whether you help a beloved fish friend make its final transition or it passes on its own, there are a number of choices as to how to dispose of it. Here are some methods that can be used. A few are rather abrupt, but most are caring and creative. A ceremonial prayer or burial service can also be included for added comfort and closure, particularly if you have children.

- Flush it down the toilet.
- Flush it down the toilet, but perform a ceremonial "Burial at Sea".
- Run it through the garbage disposal.
- Seal it in a plastic bag, then place it in the freezer and save it.
- Seal it in a plastic bag and deposit in the garbage.
- Seal it in a plastic bag or a box (plain or decorated) and bury it in the yard. A grave marker is optional.
- Immortalize it in a "Special Flower Pot Memorial".
- Perform a Viking funeral.
- Cremation.

Hopefully, your disposal experiences will be few and far between. If you would like to share any input or discuss this topic with other aquarists, go to the About Saltwater Aquariums Forums Board.

Why Did My Fish Die?

A question usually posed by beginners, even experienced saltwater aquarium hobbyists are sometimes stumped as to the reason a fish in their tank has died.

Normally, an alert aquarist can see the symptoms (frayed fins, cloudy eyes, unusual behavior, water test results) of a problem before a death occurs, but ocassionally a fish will die suddenly with no apparent reason. Barring the possibility of a fish becoming infected and dying from the West Nile Virus, below are the most probable reasons for a tank critter to die.

- Ammonia Poisoning
    - Occurs most frequently while cycling a new tank, but also happens during collection or while in transit.
    - If the tank has been fully cycled for a period of time, New Tank Syndrome may be the cause.

   - o Symptoms include cloudy eyes, frayed fins, rapid gilling and lack of appetite.
- Nitrite Poisoning
   - o Almost always occurs while cycling a tank.
   - o The symptoms are similiar to ammonia poisoning.
- Cyanide Poisoning
   - o Almost exclusively occurs in fish captured in certain areas of the South Pacific.
   - o May take a week or two to show up after collection.
   - o Other than a lack of appetite, there are few symptoms to detect.
- Poisonous Sting
   - o May be inflicted from a poisonous fish(i.e. Lionfish) or from a poisonous anemone.
   - o Difficult to diagnose.
   - o Symptoms include erratic behavior and lack of appetite.
- Other Poisons
   - o Poisons from the Neutron Bomb Boxfish, Pufferfish or Lionfish which have been released in the tank.
   - o Anything from cigarette ashes to hair spray can be sources.
- Malnutrition
   - o Most often occurs when a new fish is introduced into a tank.
   - o Lack of appetite due to disease.
   - o First symptom is a sunken belly.
- Oxygen Deprivation
   - o Normally caused by lack of vertical water movement in a tank.
   - o Initial sign is fish staying at the surface of the tank water.
- Old Age
   - o Inevitable.
   - o Difficult to diagnose.
- Physical Injuries
   - o Pretty obvious.
   - o Usually caused by other tank occupants.

- Bacterial Infection
    - o Symptoms include cloudy eyes, red areas on the body, swollen belly (internal infection).
- Bends (Diver's Disease)
    - o Rare for the hobbyist to see.
    - o Normally occurs during collection.
- Other Diseases

All About Copper

Copper Sulfate, Is it safe?

Many people have had a bad experience with, or are afraid to use Copper Sulfate on their fish because they have been improperly informed on the mixing, dosing and usage of this product.

First we will tell you what Copper does: Copper Sulfate is pretty much an all-around treatment. It is an Algaecide, meaning that it kills Algae, it is Anti-Bacterial, Anti-Fungal, works on external Protozoa such as Ich, and works on Oodinium, Sliminess of the skin, and kills certain parasites such as Cryptocaryon Irritans (Salt water Ich) and Crustacea (Argulus). It also works as a great preventative treatment for most all fish, freshwater or saltwater. We used to use it on Gill Flukes years ago, but now, the Gill Flukes have become resistant to this treatment.

How to mix up a Copper Solution properly: Stock solution= 21 grams of Copper Sulfate + 21 grams of Citric Acid Crystals to 1 pint of distilled water. Shake well. Use 1 drop per every gallon of aquarium water= .15ppm. Why use Citric Acid with Copper? Simple, it sequesters the solution to make it stable. Copper does not readily dissolve in water. The Citric Acid helps it to dissolve completely and prevent Copper levels from bouncing around.

How high can I run Copper?

Safe levels are .15ppm to .20ppm. Any higher may burn the fish and leave them with red sores on their sides. It is very important to use a Copper test kit when medicating with this product.

How do I remove Copper?

A common misconception about Copper is that it can be removed from the water with activated charcoal. It can only be removed by either doing water changes, or by using E.D.T.A. (Ethylene Diamine Tetra Acidic Acid) to chelate it out of the water. Remember folks, there is no such thing as a "Chelated Copper Solution", as "Chelated" means "Inactivated". The proper terminology would be "Sequestered Copper Solution".

Warnings before using Copper: Although Copper is safe at the correct parts per million, there are certain things that Copper cannot be use on: Saltwater Sharks, Invertebrates, and for sure... reef tanks.

Also, be careful if using Copper to kill algae. When you dose the tank, the Copper readings will disappear, because the algae will suck it all up. Then the algae dies, and releases all of the Copper back into the water.

Copper is a heavy metal ion and is considered a "Poison". Care should be used when handling this product.

Water Parameters

Saltwater Aquariums

We created this section mainly for you Saltwater Aquarists out there, to help you understand what the correct water parameters are for your tanks, and some of our importers/exporters, that have had difficulties with diseases, brought on by keeping fish in an environment that is not stable.

Yeah, Yeah, you read a book, or heard from so and so that you should keep your water a certain way. Everyone tells a different story of what they do to their tank to keep it healthy.

One of the first things that anyone should do before purchasing fish, is to ask the dealer what their water parameters are. This will ensure less stress on the fish, therefore preventing a disease to manifest.

Also, if you are living on the East Coast, or the West Coast, you must understand that the water parameters in your oceans (Atlantic & Pacific) are much different than the water parameters will be in your aquarium. Many fish come from Africa, Hawaii, Fiji and other tropical areas with much warmer water, and lower salinity levels.

Proper Salinity Levels: This is the salinity level that we use for our saltwater fish-only tanks.

This higher salinity is needed for live rock and coral in reef tanks.

Water Temperature: degrees fahrenheit. Temperatures below this will cause disease. If you keep getting Cryptocaryon (Ich), this is the reason.

Freshwater Aquariums

Freshwater aquariums are pretty easy to take care of in general, depending on the fish you decide to keep.

Water parameters for most of your common tropical fish such as livebearers, south american cichlids, goldfish, guppies etc, can be kept at the same levels.

Temperature: 76 - 80 degrees fahrenheit, pH= Neutral to Alkaline 7.0-7.6.

If you decide to keep more difficult fishes, such as Discus, water parameters are more difficult to follow.

Temperature: 84 - 86 fahrenheit, pH= Acid pH, 6.0-6.5

If you are keeping African Cichlids, they like a higher pH and harder water

Temperature: 78-80 degrees fahrenheit, pH= Alkaline pH, 8.2-8.4.

In freshwater aquariums, it is better to adapt your fish to the pH of your tap water, rather than to add chemicals every time you do a water change. If you are not breeding fishes, follow these guidelines. If you are breeding fishes, you must be careful to duplicate the water quality from where your fish came from.

Koi in the Watergarden

Koi can be included into a densely planted pond, if the area is large and deep enough. The Koi will do much better if the population is kept small and they are not confined to smaller spaces. In stocking a pond, the usual recommendation is for no more than one inch of fish per square foot of pond surface area. The water quality of well-stocked ponds should always be closely monitored to prevent ammonia levels from becoming toxic. Koi have special environmental needs and are, therefore, not really suited for life in the water garden unattended. Even though Koi may be fed sufficiently, they still require fresh greens in their diet. The submerged grasses and tender submerged lily growth often prove irresistible. Trying to curtail their appetite with daily feedings of fresh lettuce and celery leaves, may not prove to change some of their habits. The pond is a virtual smorgasbord to a Koi. It may be worthwhile to culture some small pots of leaf lettuce. You can put these in the pond if you cover the topsoil with some rocks so they won't nudge them over, and then set them in the pond for the Koi to feed on.

Like wild carp, Koi are omnivorous, and in captivity are normally fed once or twice a day. They can be offered processed food (pelletized food) that has been especially made for the Koi's nutritional requirements. Koi can be very hard on your plants because of their persistent rooting habits. They will root around looking for insects and larvae to feed on. Any valuable plants should be planted in tubs, which are then placed in the pond. Also make sure that if you are using bricks as a platform to set your plants on, to use some concrete pond paint on them to prevent them from leeching lye into the pond. This will be toxic to the fish and cause you to have extremely high pH levels.

Pond Fish

Sometimes, unfortunately... we are not blessed with good water, and must use the 8.6+ pH city tap water in our ponds. These high pH levels are not real good for many species of Koi. It may be better for you to keep some common goldfish, comet goldfish, or shubunkin's instead. These fish will grow huge just like the koi, and they will be much easier on your budget. Plus they are pretty hardy, and the shubunkin's are very colorful. Some species of catfish, tilapia and the aquarium plecostimus (algae eater) may be introduced to the pond in warmer climate zones. These fishes are scavengers and will help keep the bottom of the pond cleaner (This is no excuse not to clean your pond). Continual maintenance of your pond is crucial, and keeping a beautiful pond has proven to be too much work for many people.

Contaminated Water

Run-off water from a nearby stream, or collected rainwater may contain toxic insecticides, herbicides or fertilizers. Rainwater from metal roofs or asbestos shingles will contaminate the pond and may prove toxic to both the fish, and the plants. If the fish display signs of toxicity, execute a 50% water change and/or remove the fish to safe quarters, or a hospital tank until the water has been changed.

Acid rain may produce stress in water lilies. Immediately following a heavy rainfall, the lily leaves may show signs of burning at the edges or abrupt yellowing. A partial water change may be needed after such rainy periods, if the pH readings are lower than the neutral 7.0 range.

White foam at the waterfall entry of the pond is a sign of a high level of dissolved organic compounds. Do some partial water changes and add some Aqua Gold to the pond to handle the high organic load.

Salt on the Internet: Too many rumors about salt on the internet these days. Koi are not saltwater fish in the first place, and salting the water will not kill the bacteria in the fish's bloodstream or body cavity. Salt at high levels will eventually poison your pond and all of your fish. It would be like you waking up in the morning and drinking a big glass of saltwater for breakfast. You would end up in the hospital after a few days of doing this. Besides, use some common sense here. Salt in the water is absorbed through the fish's skin, and eventually into the muscle tissue. Kind of like those pirates stranded at sea in the old days. They knew that if you drank the sea water (saltwater), you would die. Same applies to your fish. Koi are "freshwater" fish remember?

Protozoal Parasites

The majority of the fish parasites which cause disease in fish include protozoal parasites. Typically, these parasites are present in large numbers either on the surface of the fish, within the gills, or both. When they are present in the gills, they cause problems with respiration and death will commonly occur when additional stressors are present in the aquatic environment. Protozoal parasites on the skin, fins or scales only, (i.e., not affecting the gills) usually do not result in death, unless they are accompanied by a secondary bacterial infection.

Ichthyophthirius Multifilis: This is probably the most common parasite of all fishes. The common name for this parasite and disease is "Ich" or "white spot". The mature parasite (photomicrograph) reaches approximately 1 mm in diameter and is commonly observed in the gills and/or skin as coalescing white spots, hence the common name. The trophont or mature stage of the parasite has a large "horseshoe" shaped nucleus, and the entire surface of the parasite is covered in cilia. The life cycle of this parasite is direct, but is spent, in part, off of the host. The trophont is within the epidermis of the host, until it leaves the fish, encysts (photomicrograph demonstrates mature cyst of this parasite) and divides to produce many host-seeking tomites. The tomites penetrate the skin and gills of the fish to complete the life cycle. The life cycle is temperature dependent with a shorter life cycle occurring at warmer water temperatures.

Fish with a cutaneous infection will "flash", i.e., turn over and expose their white underside, whereas fish with a gill infection will "pipe", i.e., come to the surface of the water and "breathe" through their mouth. Gill lesions include epithelial hyperplasia with the presence of mature trophonts within the gills. Cutaneous lesions also exhibit focal epidermal hyperplasia, with parasites being located beneath the hyperplastic epidermis. Photomicrograph demonstrates a single "Ich" organism with hyperplastic epidermis.

Trichodinia sp: There are three genera which form the Trichodina complex: *Trichodina, Trichodonella* and *Tripartiella,* however, all three are commonly referred to as "Trichodina". All are approximately 100 mm in diameter and have a saucer to "frisbee" shape and are ringed with cilia around its entire surface.. They have a circular arrangement of tooth-like structures (denticular ring) within the body which provides them a characteristic appearance in fresh gill and skin cytology preparations (photo). Fish with severe gill infections of trichodina will have respiratory and osmoregulatory difficulty and may "pipe" as well as "flash" if there is cutaneous involvement. Fin erosions and/or ulcerations can be observed in chronic cutaneous infections. Diagnosis of this parasitic disease is dependent upon identification of the parasite within the skin or gill cytologic preparations or histopathology.

Ambiphyra sp: These are ciliated protozoal organisms which are thought to be free-living, but have been known to parasitize fish. They are sessile organisms with a cylindrical to conical body with oral cilia and a permanent motionless equatorial ciliary fringe. They range in size from approximately 60-100 mm and adhere to the epithelium of the skin and/or gills. Disease and death of fish have been associated with chronic infections of the gills due to mechanical blockage of respiratory epithelium. Diagnosis of this parasite is dependent upon identification of this organism within the skin or gill scrapings or histopathology.

Ichthyobodo sp: This parasite is also known as Costia sp. These are obligate flagellate parasites with a direct life cycle. The free-swimming form is renniform, and approximately 10-20 mm long with two pairs of flagellae; whereas the attached form is pear-shaped and attaches to the gills and skin. Disease associated with this parasite includes increased cutaneous mucous production (hence the lay terminology of "blue slime disease"), epithelial hyperplasia of the skin and gills, ulceration and erosion of fins. This pathogen commonly causes disease in salmonid fry, resulting in high mortality. Diagnosis is dependent upon demonstration of the agent within affected fish by cytology or histopathology.

Hexamita sp: These parasites are pyriform to oval with tapering toward the posterior end. Occasionally, rounded individuals can be identified. The organisms are 6-8 mm wide and 10-12 mm long. They have three pairs of anterior flagella which are approximately one and one-half times the length of the body. The flagella originate from the blepharoplast at the anterior end of the axostyles. These organisms can reproduce by longitudinal binary fission as well as undergo schizogony within the epithelial cells of the ceca or intestine. This parasite causes disease within the gastrointestinal tract of fish and affected fish will have clinical signs related to malnutrition and emaciation. Diagnosis is dependent upon finding the parasite from cytologic scrapings of the ceca or intestinal tract or histopathology of these organs. A poorly understood parasite, thought to be *Hexamita-like* is thought to be responsible for Freshwater Hole-in-the-Head and Lateral Line Erosion (FHLLE).

Chilodonella sp: This is a motile ciliated protozoal parasite which causes disease in the skin and gills of fish. It is typically heart-shaped with the posterior end being broader and slightly notched. It measures approximately 20-40 mm in width and 30-70 mm in length and its surface is covered with cilia. There is a large macronucleus in the posterior portion of this organism and a smaller micronucleus is near or within the macronucleus. This parasite has been attributed to death of fish due to respiratory and osmoregulatory imbalances associated with severe gill parasitism (photo). Diagnosis is dependent upon demonstration of the organism within the affected organs by either cytology or histopathology.

Myxozoan Parasites

This is a very large group of parasites which can cause disease in a wide variety of fishes. They are obligate parasites of tissue (histozoic forms that reside in intercellular spaces or blood vessels that reside intracellularly) and organ cavities. Key characteristics of the Myxozoa include development of a multicellular spore, presence of polar capsules in their spores and endogenous cell cleavage in both the trophozoite and sporogony stages.

The method of transmission of myxozoans is unknown, but evidence suggests that at least some pathogenic myxozoans have an indirect life cycle. This life cycle may require the completion of two different life cycles involving a vertebrate (fish) and an invertebrate (annelid) host with each life cycle having its own sexual and asexual stages. Severe infestations by these parasites can result in disease and/or death of the host fish. Each parasite is somewhat species specific as well as organ specific.

Myxobolus Cerebralis: This parasite is known for causing "whirling disease" in salmonids. This is a chronic debilitating disease which is very common in the Western U. S. and is uncommon in the Midwestern and Eastern U. S., although recent "outbreaks" have been reported in New York. The parasite feeds on cartilage of the axial skeleton and clinical signs are related to this damage. This parasite produces spores within the cartilage which are oval to circular and approximately 7-9 mm in diameter and 6-7 mm long with a thick mucoid envelope on the posterior half of the spore (photo demonstrates spores recovered from a fish with whirling disease). Histologically, myxozoan parasites within the cartilage of the axial skeleton, with morphological characteristics of *Myxobolus cerebralis,* must be observed to confirm the diagnosis.

Aurantiactinomyxon: This is currently thought to be the causative agent of "proliferative gill disease" of catfish. This myxozoan parasite causes a rapid, severe, proliferation of the gill epithelium which results in impairment of respiration and osmoregulatory function. The intermediate host is thought to be a microscopic aquatic oligochaete worm, *Dero digitata,* which is found in the mud and sediment in the ponds of affected catfish. Diagnosis is dependent upon the presence of the parasite within swollen, clubbed proliferative gill lamellae, many of which are fractured due to the associated chondrodysplasia.

Henneguya sp: These parasites were once thought to be responsible for the "proliferative gill disease" of catfish, but are now thought to be less pathogenic, although they can cause respiratory problems if present in large numbers. They have a very characteristic paired, long, whip-like caudal processes giving them a spermatozoan-like appearance. Size of spores range from 8-24 mm in length with the caudal processes being approximately 20-45 mm long. Diagnosis is dependent upon observing the parasites within the histopathology of the gill lamellae (photomicrographs 1 and 2 demonstrate mature Henneguya cysts within the gill lamellae).

The Phylum Platyhelminthes is Composed of Three Taxonomic Classes: Turbellaria, Trematoda, and Cestoda. The Turbellaria are all free-living and have no association with fish diseases. All members of the other two classes live in close relationship with host animals. The trematodes are commonly known as *flukes,* whereas the cestodes are known as *tapeworms.*

Mongenetic Trematodes

This is a group of trematodes which complete their entire life cycle on the host. The adults attach to the host by a *haptor* or *opishaptor* which is a specially adapted structure on the posterior end of the parasite. This organ has hooks which allow the parasite to attach firmly to the host fish. These parasites usually cause minimal damage to fish, but will infest the skin, fin and gills of pond fishes. Severe infestations may be responsible for poor respiration and/or emaciation. The two most common monogenetic trematodes include: *Dactylogyrus* and *Gyrodactylus.*

Dactylogyrus sp: This parasite is approximately 0.2 to 0.5 mm in length, reaching a maximum length of 2.0 mm. It has seven pairs of marginal hooks and usually one pair of median hooks on the opishaptor. The dactylogrids have two to four pigmented spots (known as "eyes" or "eye spots") in the anterior fourth of the body. All dactylogrids are oviparous with no uterus (photo).

Gyrodactylus sp: This parasite is smaller, rarely reaching a maximum length of over 0.4 mm. All species are viviparous with one to three daughter generations being observable in the V-shaped uterus. This parasite is more commonly found on the skin and less commonly present in the gills, although severe infestation will have both organs affected.

Acanthocephala

This group of parasites is comprised of worms with an anterior proboscis covered with many hooks. These parasites are often referred to as "thorny-headed worms". The body is composed of a presoma (proboscis and associated structures) and a cylindrical trunk. The intestinal tract of the affected fish may contain a blood-tinged fluid, and histologically, the proboscis of these parasites will be firmly attached to the intestinal mucosa. Photomicrograph demonstrates the proboscis (top) of an acanthocephala parasite which has burrowed into the intestinal mucosa (bottom) of an affected fish. Affected fish will exhibit emaciation, lethargy, anemia, and possibly death with a marked infection. We have observed salmonids from the Great Lakes with marked infestation of this parasite with literally hundreds of these parasites embedded in the intestinal mucosa (photo).

Leeches and Copepods

Although not a common problem, occasionally, fish will be observed infected with either leeches or copepods. Leeches have long, slender flexible bodies and actively swim for an attack on their prey. Skin and underlying soft tissues are damaged and allow blood to flow into the leeches digestive tract. Leeches are not host-specific, and the damage to the skin and gills is dependent upon the number of leeches present at any time. Small fishes can be seriously injured or die due to excessive leech infestation.

Copepods include fish lice or "anchor worms". The more common fish lice include *Lepeophtheirus* and *Caligus,* and *Argulus.* The most common genus of anchor worms includes *Lernae sp (photo demonstrates a typical Lernae sp. anchor worm).* All of these are external parasites which affect the fish by imbibing blood from the host fish and causing localized skin and soft tissue damage. They may also allow for secondary bacterial infection of the skin or musculature which may ultimately cause the demise of the fish.

5

Common Disorders

Ichthyophthirius Multifiliis

Ichthyophthirius multifiliis (commonly known as freshwater white spot disease, freshwater ich, or freshwater ick) is a common disease of freshwater fish. It is caused by the protozoa *Ichtyopthirius*. Ich is one of the most common and persistent diseases. The protozoan is an ectoparasite. White nodules that look like white grains of salt or sugar of up to 1 mm appear on the body, fins and gills. Each white spot is an encysted parasite.

Description

I. multifiliis is one of the most prevalent protozoan parasites of fish and is an important pathogen of ornamental and farm-raised food fish species when reared under intensive conditions. Wild fish populations are also susceptible and outbreaks are occasionally seen. There are few aquarists that have not met it on one or more occasions.

The ich protozoa goes though the following life stages:

- *Feeding stage :* The ich trophozoite (a protozoan in active stage of life) feeds in a nodule formed in the skin or gill epithelium.
- After it feeds within the skin or gills, the trophozoite falls off and enters an encapsulated dividing stage (tomont). The tomont adheres to plants, nets, gravel or other ornamental objects in the aquarium.
- The tomont divides up to 10 times by binary fission, producing infective theronts, thus dividing rapidly and attacking the fish.

This life cycle is highly dependent on water temperature, and the entire life cycle takes from approximately 7 days at 25 °C (77 °F) to

8 weeks at 6 °C (43 °F). Marine ich is a similar disease caused by a different ciliate, *Cryptocaryon irritans*.

Predisposing Factors

There is no dormant stage in the lifecycle. Ich does not lie in wait for a weakened fish to infect. However, any factor that reduces immunity like changes in water temperature and quality may, in a subclinically infected fish, accelerate an outbreak of Ich. The presence of ammonia, nitrite and high levels of nitrate in water does not in itself cause clinical cases of Ich. However, poor water quality will stress fish, allow an outbreak to spread rapidly and increase mortality rates in infected fish.

Diagnosis

Typical behaviours of clinically infected fish include:

- Anorexia (loss of appetite, refusing all food, with consequential wasting)
- Rapid breathing
- Hiding abnormally/ not schooling
- Resting on the bottom
- Flashing
- Rubbing and scratching against objects

A subclinically infected fish will not show any of these signs. For example, a healthy fish with a newly attached trophozoite will not yet have clinical disease. The trophozoite will not become visible to the naked eye until it has fed on the fish and grown to one or two millimetres. A trophozoite attached to the gills usually is not readily seen. A subclinically infected fish may initially only have a single trophozoite.

Skin

Visible Ich lesions are usually seen as one or several characteristic white spots on the body or fins of the fish. The white spots are single cells called trophozoites or trophonts, which feed on the tissues of the host and may grow to 1 mm in diameter. A smear should show ciliates if white spot is present.

Eyes

The eye becomes cloudy almost to the point of whiteness and the fish lose vision. The causes behind this disease can vary. An increase in parasites in the aquarium is the most common cause but severe stress, old age, or malnutrition can all lead to this condition. Treating this condition requires an investigation of water quality. Once the water quality is high enough, the fish will usually recover by themselves within 1–2 weeks. Thus, it is advisable to wait for 1–2 weeks before administering antibiotics.

Gills

Gill infection will cause breathing at the surface and fast respiration. Gill examination may show numbers of such white spots. Wet mount of a Gill Biopsy may show myself, the mutifiliis trophozoites.

Treatment

Unfortunately, an efficient prevention of the disease by vaccination is not possible, although several studies identified potential vaccine candidate proteins, i.e. i-antigens, of the parasite. Any treatment method must take into account the species of fish (some will not tolerate certain medications), how high the infection rate is, and the size and type of environment. If it is detected before it becomes too serious, a number of different treatments can be applied. Only the free-swimming stage of the parasite is susceptible to treatment; neither the trophonts under the epithelium nor the tomont cysts can be killed.

Heat Treatment

Heat treatment can be highly effective, and it can be combined with other treatments. However, it can only be used on fish that can tolerate high water temperatures, and is unsuitable for cold water fish like koi and goldfish, but even in those cases, a higher water temperature will accelerate the life-cycle of the parasite, allowing other treatments to take effect sooner.

Chlorine

For treating koi & goldfish, chlorine, in the form of tapwater, is very effective in removing not only the threadlike parasites, but eventually the persistent cysts. Thread like infestations on fish will disappear overnight, cysts will take a couple of weeks and possibly a couple of water changes to eliminate. Aquarium lighting is used to detect the presence of parasites, as the filament like threads fluoresce at these light frequencies.

Salt

One method of treatment for ich consists of adding aquarium salt until a specific gravity of 1.002 g/cm^3 is achieved, as the parasites are less tolerant of salt than fish. This is not practical in ponds because even a light salt solution of 0.01% (100 mg/L; pure water at 4 °C/ 39 °F), would require large quantities of salt. Fish can be dipped in a 0.3% (3g/L; pure water at 4 °C) solution for thirty seconds to several minutes, or they can be treated in a prolonged bath at a lower concentration (0.05% = 500 mg (0,5g)/L; pure water at 4 °C).

Chemical Treatments

Chemical treatments include formalin, malachite green, chelated copper, copper sulfate, potassium permanganate and quinine. There are also a large number of proprietary treatments available for the treatment of white spot, and the related Oodinium (velvet disease). All treatments target the free-living theronts and tomonts, which only survive about two to three days in the absence of a host fish.

Prognosis

When Ich is diagnosed early, effective treatment is used, and stresses are minimised, mortality rates can be low. However, if the infection is at an advanced stage, treatment protocols are not followed, and the fish are stressed, higher death rates will occur. When a fish has had Ich eradicated, it may develop partial resistance to reinfection. Partially treated fish may initially harbour low numbers of unseen trophozoites, often in the gills. This subclinical carrier will cause another outbreak weeks later, most likely when stresses occur or uninfected fish are introduced to the aquarium.

Cryptocaryon

Cryptocaryon irritans (also known as marine white spot disease or marine ich) is a species of ciliate protozoa that parasitizes marine fish, and is one of the most common causes of disease in marine aquaria.

Taxonomy

Cryptocaryon irritans was originally classified as *Ichthyophthirius marinus*, but it is not closely related to the other species. It belongs to the class Prostomatea, but beyond that its placement is still uncertain.

Clinical

The symptoms and life-cycle are generally similar to those of *Ichthyophthirius* in freshwater fish, including white spots, on account of which *Cryptocaryon* is usually called marine ich. However, *Cryptocaryon* can spend a much longer time encysted.

Useful treatments of *Cryptocaryon irritans* are copper solutions, formalin solutions and quinine based drugs (such as Chloroquine Phosphate and Quinine Diphosphate).

Infections can be extremely difficult to treat because of other creatures, such as corals and other invertebrates, which will not survive standard treatments. Ideally fish with *Cryptocaryon* are quarantined in a hospital tank, where they can be treated with a copper salt or using hyposalinity. The display tank needs to be kept clear of fish for 6-9 weeks, the longer the better. This gives time for the encysted tomonts to release infectious theronts, which die within 24-48 hours when they cannot find a host.

Kudoa Thyrsites

Kudoa thrysites is a myxosporean parasite of marine fishes. It has a worldwide distribution, and infects a wide range of host species. This parasite is responsible for causing economic losses to the fisheries sector, by causing post-mortem "myoliquefaction", a softening of the flesh to such an extent that the fish becomes unmarketable. It is not infective to humans.

Taxonomy

The spores of *K. thyrsites* are stellate in shape, with 4 valves and 4 polar capsules. Upon infection by the actinosporean stage the sporoplasm migrates to a muscle fibre where it forms a pseudocyst. Within these pseudocysts are the developing spore stages. Comparison of 18S rDNA sequences of *Kudoa* species and other myxozoan species to determine their relationships. They show that *Kudoa* species are distinct from other myxozoans analyzed (*Myxidium* sp., *Myxobolus* sp., and *Henneguya zschokkei*). *Kudoa thyrsites* is an interesting member of this group in that apparently has very broad host specificity, infecting many fish species around the world.

Pathology

Members of the genus Kudoa primarily infect muscle tissue of marine fishes, where they form nodules or pseudocysts containing a great number of individual spores. In lighter infections these pseudocysts are isolated from the fish's immune system within the muscle fibre. More intense infections can result in severe inflammation surrounding infected muscle fibres. Although apparently asymptomatic in all but heavy infections, they are associated with post-mortem degeneration of the tissue. This softening of flesh is most likely a result the release of proteolytic enzymes by the parasite. This causes losses to both aquaculture operations, for instance, where salmon are being reared in "sea-pens", and to capture fisheries. Losses are both direct, through the degradation of fish products, and indirectly, through the perception of the consumer that fish from a particular area are of a lower quality. The intensity of *K. thyrsites* infection is positively correlated with the severity of flesh softening in Atlantic salmon fillets. Softening of flesh always occurred with heavily infected fillets, while lightly infected fillets showed no softening. Prevention and/or control of *K. thyrsites* infections is problematic especially in open water netpens. Currently there are no available treatments. One approach to control may be to disrupt the life cycle in some way thereby minimizing the likelihood of infection.

Table: *Distribution and species infected by* Kudoa thrysites

Location	***Species***	***Common name***
North America	Merluccius productus	*Pacific hake*
	Oncorhynchus *spp.*	*Pacific salmon*
	Icelinus filamentosus	*Threadfin sculpin*
	Ophiodon elongatus	*Lingcod*
	Aulorhychus flavidus	*Tube-snout*
	Salmo salar	*Atlantic Salmon*
	Reinhardtius stomias	*Arrowtooth Flounder*
	Eopsetta jordani	*Petrale sole*
	Hippoglossus stenolepis	*Pacific halibut*
	Microstomus pacificus	*Dover sole*
	Lepidopsetta bilineatus	*Rock sole*
	Platichthys stellatus	*Starry flounder*
	Parophrys vetula	*English sole*
	Theragra chalcogramma	*Alaskan pollock*
	Merluccius capensis	*Cape hake*
Australia	Engraulis australis	*Australian anchovy*
	Engraulis japonicus	*Japanese anchovy*
	Sardinella lemuru	*Bali sardinella*
	Sardinops neopilchardus	*Australian pilchard*
	Spratelloides delicatulus	*Blue sprat*
	Coryphaena hippurus	*Mahi Mahi*
South Africa	Sardinops ocellatus	*South African pilchard*
	Thyrsites atun	*Snook*

Contd...

Location	**Species**	**Common name**
Chile	Paralichthys adspersus	*Fine flounder*
Japan	Cypselurus *sp.*	*Flying fish*
Ireland	Salmo salar	*Atlantic Salmon*
	Clupea harengus	*Herring*
Spain	Salmo salar	*Atlantic Salmon*
United Kingdom	Scomber scomber	*Mackerel*
	Salmo trutta	*Street trout*

Life Cycle

The myxosporeans have been shown to have complex life cycles using more than one host. Usually a fish and an oligochaete or polychaete worm, and in one case a bryozoan. The life cycle of *K. thrysites* is poorly understood. It has been hypothesized that *K. thyrsites* has an indirect life cycle involving some marine invertebrate. Experiments have shown direct transmission of the marine myxosporean *Myxidium leei* in sea bream. However, direct transmission of K. thyrsites failed when naive fish were fed fresh myxospores. If K. thyrsites does have an indirect life cycle, the intermediate host has yet to be identified.

Lernaeocera Branchialis

Lernaeocera branchialis, sometimes called cod worm, is a parasite of marine fish, found mainly in the North Atlantic. It is a marine copepod which starts life as a small pelagic crustacean larvae. It is among the largest of copepods, ranging in size from 2–3 millimetres when it matures as a copepodid larva to more than 40 millimetres (1.6 in) as an adult.

Lernaeocera branchialis is ectoparasitic, which means it is a parasite that lives primarily on the surface of its hosts. It has many life stages, some of which are motile and some of which are sessile. It goes through two parasitic stages, one where it parasites as a secondary host a flounder or lumpsucker, and another stage where it parasites as a primary host a cod or other fishes of the cod family (gadoids). It is a pathogen that negatively impacts the commercial fishing and mariculture of cod-like fish.

Life Stages

The life-cycle of a cod worm involves a complex progression of life stages, including two successive hosts. It comprises "two free-swimming nauplius stages, one infective copepodid stage, four chalimus stages and the adult copepod, each separated by a moult". The cycle begins with the females laying eggs which hatch into a nauplius, the usual early larval stage of crustaceans. This nauplius moults about 10 minutes after hatching to produce nauplius II, and 48 hours later, nauplius II moults to a copepodid stage. At this point the copepodid is pelagic and free-swimming with an average length of about 0.5 mm.

The next stage is finding a secondary or intermediate host, a demersal fish like a flounder or lumpfish which is often stationary and therefore easy to catch. The copepodid have only a day to find such a fish and attach themselves to its gills.

When they locate such a fish, they capture it with grasping hooks at the front of their body. They penetrate the fish with a thin filament which they use to suck its blood. The nourished cod worms then progress via four moults from the naupliar stage to the mature chalimus stage. At this point the males transfer sperm to the females. Both sexes develop swimming setae, detach from the flounder or lumpfish and again swim freely as pelagic organisms.

The female worm still resembles a copepod and is 2 to 3 mm long. She now undergoes another pelagic quest, searching this time for a definitive or primary host. With her fertilised eggs, she looks for a cod or a fish belonging to the same family as cod, such as a haddock or whiting.

When she locates one the worm enters the gill chamber. There she clings to the gills and metamorphoses into a plump, sinusoidal, wormlike body, with a coiled mass of egg strings at her rear. These bodies are mostly about 20 mm long, but can measure up to 50 mm. The front part of the worm's body penetrates the body of the cod until it enters the rear bulb of the host's heart. There, firmly rooted in the cod's circulatory system, the front part of the parasite develops in the shape of antlers or branches on a tree, reaching into the main artery. In this way, while safely tucked beneath the cod's gill cover, the worm feeds from one end on cod blood while it pumps new offspring out the other end.

Behaviour

It is not known how *L. branchialis* searches for its fish hosts, but it probably uses chemoreceptors and mechanoreceptors, and follows physical clues in the water column such as those provided by haloclines and thermoclines.

Effects on Fisheries

The most serious parasitic crustaceans among fish in general are sea lice. However, *L. branchialis* is probably the most serious parasitic crustacean among cod. Infestation reduces the efficiency with which food can be utilised, delaying the development of the gonads. Up to 30% loss in weight can occur, with increases in mortality because of open lesions with loss of blood, and possibly occlusion of vessels or aorta. These can have commercial impacts on wild fisheries, making cod-like fishes more expensive to market. Gadoids, particularly cod, are emerging marine aquaculture species in some North Atlantic countries. *L. branchialis* present potential problems for their successful mariculture.

Myxobolus Cerebralis

Myxobolus cerebralis is a myxosporean parasite of salmonids (salmon, trout, and their allies) that causes whirling disease in farmed salmon and trout and also in wild fish populations. It was first described in rainbow trout in Germany a century ago, but its range has spread and it has appeared in most of Europe (including Russia), the United States, South Africa and other countries. In the 1980s, it was discovered that *M. cerebralis* needs to infect a tubificid oligochaete (a kind of segmented worm) to complete its life-cycle. The parasite infects its hosts with its cells after piercing them with polar filaments ejected from nematocyst-like capsules.

Whirling disease afflicts juvenile fish (fingerlings and fry) and causes skeletal deformation and neurological damage. Fish "whirl" forward in an awkward corkscrew-like pattern instead of swimming normally, find feeding difficult, and are more vulnerable to predators. The mortality rate is high for fingerlings, up to 90% of infected populations, and those that do survive are deformed by the parasite residing in their cartilage and bone. They act as a reservoir for the parasite, which is released into water following the fish's death. *M. cerebralis* is one of the most economically important myxozoans in fish as well as one of the most pathogenic. It was the first myxosporean whose pathology and symptoms were described scientifically. The parasite is not transmissible to humans.

Taxonomy

The taxonomy and naming of both *M. cerebralis* and of myxozoans in general have complicated histories. It was originally thought that this parasite infected fish brains (hence the specific epithet *cerebralis*), however it quickly became apparent that while it can be found in the nervous system, it primarily infects cartilage and skeletal tissue. Attempts to change the name to *Myxobolus chondrophagus*, which would more accurately describe the organism, failed because of nomenclature rules.

Later, it became apparent that organisms previously called *Triactinomyxon dubium* and *T. gyrosalmo* (class Actinosporea) were in fact triactinomyxon stages of *M. cerebralis,* whose life cycle was expanded to include the triactinomyxon stage. Similarly, other actinosporeans were folded into the life cycles of various myxosporeans.

Today, the myxozoans, previously thought to be multicellular protozoans are considered animals by many scientists, though their status has not officially changed. Recent molecular studies suggest that they are related to Bilateria or Cnidaria, with Cnidaria being closer morphologically because both groups have extrusive filaments, but with Bilateria being somewhat closer in some genetic studies.

Morphology

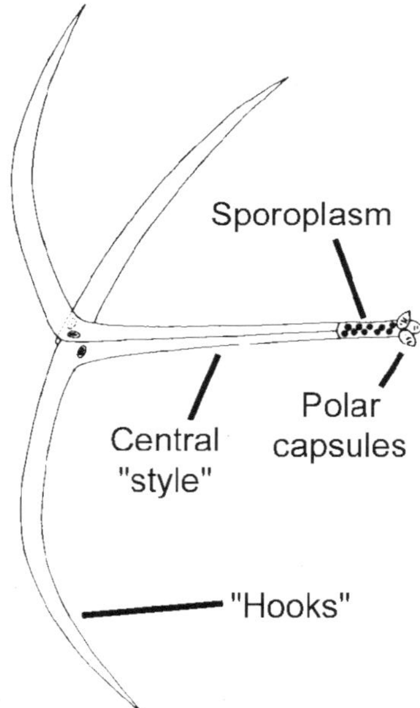

Figure: *Diagram of the structure of a triactionmyxon stage spore of* Myxobolus cerebralis

M. cerebralis has many diverse stages ranging from single cells to relatively large spores, not all of which have been studied in detail.

Triactinomyxon Stage

The stages that infect fish, called triactinomyxon spores, are made of a single style that is about 150 micrometers (μm) long and three processes or "tails" that are each about 200 micrometers long. A sporoplasm packet at the end of the style contains 64 germ cells surrounded by a cellular envelope. There are also three polar capsules, each of which contains a coiled polar filament between 170 and 180 μm long. Polar filaments in both this stage and in the myxospore stage rapidly shoot into the body of the host, creating an opening through which the sporoplasm can enter.

Sporoplasm Stage

Upon contact with fish hosts and firing of the polar capsules, the sporoplasm contained within the central style of the triactinomyxon migrates into the epithelium or gut lining. Firstly, this sporoplasm undergoes mitosis to produce more amoeboid cells, which migrate into deeper tissue layers, in order to reach the cerebral cartilage.

Myxosporean Stage

Myxospores, which develop from sporogonic cell stages inside fish hosts, are lenticular. They have a diameter of about 10 micrometers and are made of six cells. Two of these cells form polar capsules, two merge to form a binucleate sporoplasm, and two form protective valves. Myxospores are infective to oligochaetes, and are found among the remains of digested fish cartilage. They are often difficult to distinguish from related species because of morphological similarities across genera. Though *M. cerebralis* is the only myxosporean ever found in salmonid cartilage, other visually similar species may be present in the skin, nervous system, or muscle.

Life Cycle

Myxobolus cerebralis has a two-host life-cycle involving a salmonid fish and a tubificid oligochaete. So far, the only worm known to be susceptible to *M. cerebralis* infection is *Tubifex tubifex*, though what scientists currently call *T. tubifex* may in fact be more than one species. First, myxospores are ingested by tubificid worms. In the gut lumen of the worm, the spores extrude their polar capsules and attach to the gut epithelium by polar filaments. The shell valves then open along the suture line and the binucleate germ cell penetrates between the intestinal epithelial cells of the worm. This cell multiplies, producing many amoeboid cells by an asexual cell fission process called merogony. As a result of the multiplication process, the intercellular space of the epithelial cells in more than 10 neighbouring worm segments may become infected.

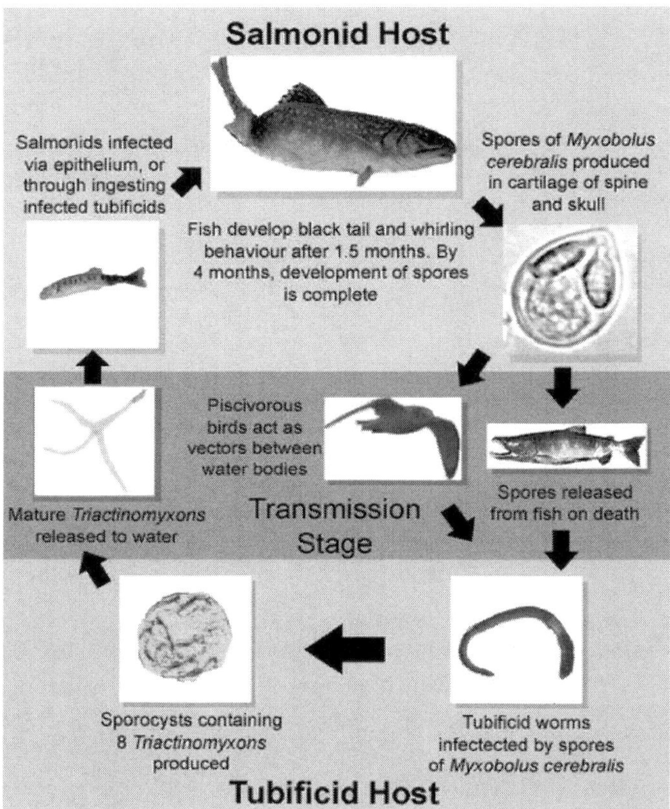

Figure: *Life cycle of* M. cerebralis. *Click to enlarge.*

Around 60–90 days post-infection, sexual cell stages of the parasite undergo sporogenesis, and develop into pansporocysts, each of which contains eight triactinomyxon-stage spores. These spores are released from the oligochaete anus into the water.

Alternatively, a fish can become infected by eating an infected oligochaete. Infected tubificids can release triactinomyxons for at least 1 year. The triactinomyxon spores swim through the water to infect a salmonid through the skin. Penetration of the fish by these spores takes only a few seconds. Within five minutes, a sac of germ cells called a sporoplasm has entered the fish epidermis, and within a few hours, the sporoplasm splits into individual cells that will spread through the fish.

Within the fish, there are both intracellular and extracellular stages that reproduce in its cartilage by asexual endogeny, meaning that new cells grow from within old cells. The final stage within fish is the myxospore, which is formed by sporogony. They are released into the environment when the fish decomposes or is eaten. Some recent research indicates that some fish may expel viable myxospores while still alive.

Myxospores are extremely tough: "it was shown that *Myxobolus cerebralis* spores can tolerate freezing at "20°C for at least 3 months, aging in mud at 13°C for at least 5 months, and passage through the guts of northern pike *Esox lucius* or mallards *Anas platyrhynchos* without loss of infectivity" to worms. Triactinomyxons are much shorter lived, surviving 34 days or less, depending on temperature.

Pathology

Figure: *Skeletal deformation in a mature Brook trout caused by* M. cerebralis *infection. Photo by Dr. Thomas L. Wellborn, Jr.*

M. cerebralis infections have been reported from a wide range of salmonid species: 8 species of "Atlantic" salmonids, *Salmo*; 4 species of "Pacific" salmonids, *Oncorhynchus*; 4 species of Char, *Salvelinus*; the Grayling, *Thymallus thymallus*; and the Huchen, *Hucho hucho*. *M. cerebralis* causes damage to its fish hosts through attachment of triactinomyxon spores and the migrations of various stages through tissues and along nerves, as well as by digesting cartilage. The fish's tail may darken, but aside from lesions on cartilage, internal organs generally appear healthy. Other symptoms include skeletal deformities and "whirling" behavior (tail-chasing) in young fish, which was thought to have been caused by a loss of equilibrium, but is actually caused by damage to the spinal cord and lower brain stem. Experiments have shown that fish can kill *Myxobolus* in their skin (possibly using antibodies), but that the fish do not attack the parasites once they have migrated to the central nervous system. This response varies from species to species.

In *T. tubifex*, the release of triactinomyxon spores from the intestinal wall damages the worm's mucosa; this may happen thousands of times in a single worm, and it is believed that this can impair nutrient absorption. Also, infected worms have lower body mass and may be discolored. Spores are released from the worm almost exclusively when the temperature is between 10°C and 15°C, so fish in warmer or cooler waters are less likely to be infected, and infection rates vary seasonally.

Susceptibility

Fish size, age, concentration of triactinomyxon spores, and water temperature all affect infection rates in fish, as does the species of the fish in question. The disease has the biggest impact on fish less than five months old because their skeleton has not ossified. This makes young fish more susceptible to deformities and provides *M. cerebralis* more cartilage on which to feed. In one study of seven species of many strains, brook trout and rainbow trout (except one strain) were far more heavily affected by *M. cerebralis* after two hours of exposure than other species were, while bull trout, Chinook salmon, brown trout, and arctic grayling were least severely affected. While brown trout may harbor the parasite, they typically do not show any symptoms, and this species may have been *M. cerebralis'* original host. This lack of symptoms in brown trout meant that the parasite was not discovered until after nonnative rainbow trout were introduced in Europe. The susceptibility of various salmonids is listed in Salmonid susceptibility to whirling disease.

Diagnosis

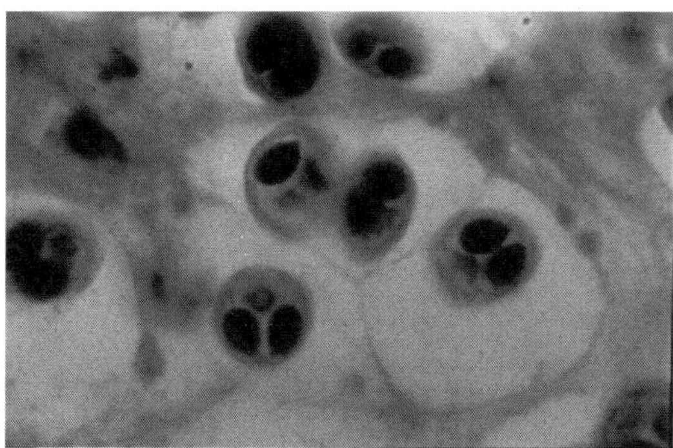

Figure: *The normally uniform trout cartilage is scarred with lesions in which* M. cerebralis *spores develop, weakening and deforming the connective tissues.*

Moderate or heavy clinical infection of fish with whirling disease can be presumptively diagnosed on the basis of changes in behavior and appearance about 35 to 80 days after initial infection, though "injury or deficiency in dietary tryptophan and ascorbic acid can evoke similar signs," so conclusive diagnosis may require finding myxospores in the fish's cartilage. In heavy infections, examining cartilage microscopically may be all that is needed to find spores. In less severe infections the most common test involves digestion of the cranial cartilage with the proteases pepsin and trypsin (the pepsin-trypsin digest—PTD) before looking for spores. The head and other tissues can be further examined using histopathology to confirm that the location and morphology of the spores matches what is known for *M. cerebralis*. Serological identification of spores in tissue sections using an antibody raised against the spores is also possible. Parasite identity can also be confirmed using the polymerase chain reaction to amplify the 415 base pair 18S rRNA gene from *M. cerebralis*. Fish should be screened at the life stage most susceptible to the parasite. Routine screening using these techniques is carried out in countries where the parasite occurs and in countries like Australia and Canada that are not known to have the parasite but where its introduction could threaten local fish.

Impact

Although originally a mild pathogen of *Salmo trutta* in central Europe and other salmonids in north east Asia, the spread of the Rainbow trout (*Oncorhynchus mykiss*) has greatly increased the impact of this parasite. Having no innate immunity to *M. cerebralis*, rainbow trout are particularly susceptible, and can release so many spores that even more resistant species in the same area, like *S. trutta*, can become overloaded with parasites and incur 80%–90% mortalities. Where *M. cerebralis* has become well-established, it has caused decline or even elimination of whole cohorts of fish.

Impact in Europe

The impact of *M. cerebralis* in Europe is somewhat lessened by the fact that the species is endemic to this region, giving native fish stocks a degree of immunity. Rainbow trout, the most susceptible species to this parasite, are not native to Europe; successfully reproducing feral populations are rare, so there are few wild rainbow trout that are young enough to be susceptible to infection. On the other hand, they are widely reared for restocking sport-fishing waters and for aquaculture, where this parasite has its greatest impact.

Hatching and rearing methods designed to prevent infection of Rainbow trout fry have proved successful in Europe. These techniques include hatching eggs in spore-free water and rearing fry to the "ossification" stage in tanks or raceways. These methods give particular attention to the quality of water sources in order to guard against spore introduction during water exchanges. Fry are moved to earthen ponds only when they are considered to be clinically resistant to the parasite, after skeletal ossification occurs.

Impact in New Zealand

M. cerebralis was first found in New Zealand in 1971. The parasite has only been found in rivers in the South Island, away from the most important aquaculture sites. Additionally, salmonid species commercially aquacultured in New Zealand have low susceptibility to whirling disease, and the parasite has also not been shown to affect native salmonids. An important indirect effect of the parasites presence is quarantine restriction placed on exports of salmon products to Australia.

Impact in the United States

M. cerebralis was first recorded in North America in 1956 in Pennsylvania, having been introduced via infected trout imported from Europe, and has spread steadily south and westwards. Until the 1990s, whirling disease was considered a manageable problem affecting rainbow trout in hatcheries. However, it has recently become established in natural waters of the Rocky Mountain states (Colorado, Wyoming, Utah, Montana, Idaho, New Mexico) where it is causing heavy mortalities in several sportfishing rivers.

Some streams in the western United States have lost 90% of their trout. In addition, whirling disease threatens recreational fishing, which is important for the tourism industry, a key component of the economies of some U.S. western states. For example, "the Montana Whirling Disease Task Force estimated that trout fishing generated US $300,000,000 in recreational expenditures in Montana alone". Making matters worse, some of the fishes that *M. cerebralis* infects (bull trout, cutthroat trout, and steelhead) are already threatened or endangered, and the parasite could worsen their already precarious situations. For reasons that are poorly understood, but probably have to do with environmental conditions, the impact on infected fish has been greatest in Colorado and Montana and least in California, Michigan, and New York.

Prevention and Control

Some biologists have attempted to disarm triactinomyxon spores by making them fire prematurely. In the laboratory, only extreme acidity or basicity, moderate to high concentrations of salts, or electrical current caused premature filament discharge; neurochemicals, cnidarian chemosensitizers, and trout mucous were ineffective, as were anesthetized or dead fish. If spores could be disarmed, they would be unable to infect fish, but it is unclear whether any of the methods that worked in the laboratory could be employed in the wild.

Some strains of fish are more resistant than others, even within species; using resistant strains may help reduce the incidence and severity of whirling disease in aquaculture. There is also some circumstantial evidence that fish populations can develop resistance to the disease over time. Additionally, aquaculturists may avoid *M. cerebralis* infections by not using earthen ponds for raising young fish; this keeps them away from possibly infected tubificids and makes it easier to eliminate spores and oligochaetes through filtration, chlorination, and ultraviolet bombardment. To minimise tubificid population techniques include periodic disinfection of the hatchery or aquaculture ponds and the rearing of small trout indoors in pathogen-free water. Smooth-faced concrete or plastic-lined raceways that are kept clean and free of contaminated water keep aquaculture facilities free of the disease.

Lastly, some drugs such as furazolidone, furoxone, benomyl, fumagillin, proguanil and clamoxyquine have been shown to impede spore development, which reduces infection rates. For example, one study showed that feeding Fumagillin to *Oncorhynchus mykiss* reduced the number of infected fish from between 73% and 100% to between 10% and 20%. Unfortunately, this treatment is considered unsuitable for wild trout populations, and no drug treatment has ever been shown to be effective in the studies required for United States Food and Drug Administration approval.

Recreational and sports fishers can help to prevent the spread of the parasite in a number of ways. Cleaning fishing equipment between fishing trips and never transporting fish from one body of water to another should protect against cross contamination of waterways. Spores are particularly persistent in felt soled wading shoes, which can be treated with 10% chlorine bleach and water for at least 15 minutes and then rinsed thoroughly. Fish bones or entrails should never be disposed of in any body of water, since spores from the carcass will be released into the waterway. Salmon and trout should not be used as bait.

Nanophyetus Salmincola

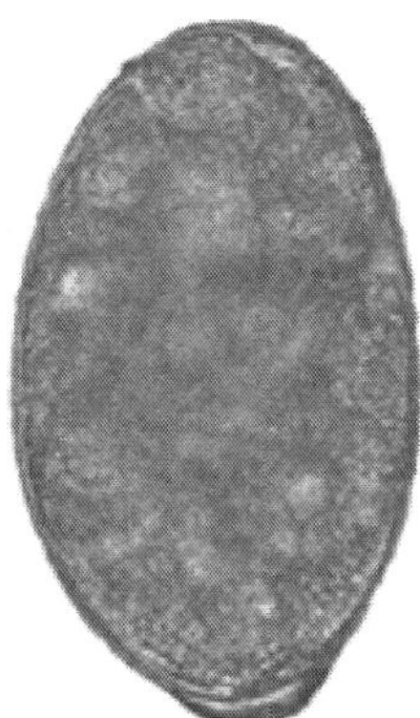

Figure: *An egg of* Nanophyetus salmincola

Nanophyetus salmincola may be the most common trematode endemic to the United States. In particular, the parasite is a food-borne intestinal trematode prevalent in the coast of the Pacific Northwest.

The life cycle of the *N. salmincola* requires three hosts. The first intermediate host is an *Oxytrema silicula* stream snail. The second intermediate host is a salmonid fish, though some non-salmonid fishes also play a role. Lastly, the definitive host is most commonly a canid, though many other mammals are also definitive hosts, including humans. Transmission of *N. salmincola* to the definitive host occurs upon ingestion of parasite-infected fish.

The parasite is most known for its association with "salmon poisoning disease", which, left untreated, proves to be fatal to dogs and other canids. However, canids are affected by the *Neorickettsia helminthoeca* bacteria, for which *N. salmincola* acts as a vector, and not by the parasite itself.

Very few known cases of naturally acquired human infection with *N. salmincola* are found in the literature, though it is likely that many cases are unreported, since most people are asymptomatic, or symptomatic with non-specific symptoms like gastrointestinal discomfort. Disease caused by *N. salmincola*, or nanophyetiasis, is easily preventable by thoroughly cooking fish before consumption. There are no known cases of human infection by the *Rickettsia* bacteria carried by *N. salmincola*. A subspecific parasite, *Nanophyetus schikhobalowi*, is endemic to Siberia, where human cases of nanophyetiasis have been reported in scientific literature since 1931.

Agent (classification and taxonomy)

Kingdom: Animalia Phylum: Platyhelminthes Class: Trematoda Order: Digenea Family: Troglotrematidae Genus: Nanophyetus Species: Salmincola

Synonyms

Nanophyetus salmincola (Chapin), *Troglotrema salmincola*, and *Nanophyetus schikhobalowi* (Russian form)

History of Discovery

The first record of salmon poisoning disease (SPD) was reported in northwestern Oregon in 1814 when a writer for Henry's Astoria Journal noted the death of dogs after consumption of raw salmon., At first, investigators believed that SPD was caused by poisonous blood in the ingested fish. In 1911, small white cysts were observed in the kidneys of disease-causing salmon and trout, but the cysts were mistakenly identified as amebae. Small trematodes in the intestines of dogs that died after eating infected salmon were finally found in 1925 and the cysts present in the salmon were correctly identified as intermediate stages of the trematode. In an experimental follow-up study, researchers showed that the small intestinal parasite did in fact cause SPD in dogs, and that the cysts did develop into the adult worm found in the intestine.

The trematode was first named by Chapin as *Nanophyes salmincola* in 1926, as a member of the family Heterophyidae. Upon further examination of the morphology, Chapin reassigned the trematode to the family Troglotrematidae and renamed the parasite *Nanophyetus salmincola*, since *Nanophyes* was already taken. Discussions regarding the correctness of classification of the parasite continued as the trematode received further scientific attention and its morphology and behavior was further scrutinized. Ultimately, *Nanophyetus salmincola* was agreed upon, though *Troglotrema salmincola* remains a synonym.

In 1931, Skrjabin and Podjapolskaja describe a similar parasite, *Nanophyetus schikhobalowi*, which was endemic to East Siberia. Argument regarding whether or not *N. schikhobalowi* and *N. salmincola* were the same or different species recurred until 1966 when the two were granted subspecific status in order to reflect their biological and geographic differences, but little significant morphological differences.,, Since its discovery, *N. schikhobalowi* has been known to naturally infect humans and research reveals surveys indicating rates of infection in endemic Siberian villages of up to 98%.

In contrast, *N. salmincola* was not recognized to be a source of an infection until a researcher purposefully infected himself in a scientific experiment in 1958. Besides Philip experimentally infecting himself with the North American *N. salmincola*, the first naturally acquired human intestinal infection cases were observed between September 1974 and October 1985. The study revealed 10 patients who presented with positive *N. salmincola* stool samples and either gastrointestinal complaints or otherwise unexplainable peripheral blood eosinophilia. 7 patients recalled ingestion of undercooked or raw fish. Of those who were not given effective treatment, symptoms and/ or eggs in stools persisted for 2 or more months before spontaneously resolving. It was hypothesized that the movement, attachment, and irritation of the adult worms in the small intestine mucosa was the likely cause of gastrointestinal symptoms and peripheral eosinophilia.

Two years after the first 10 cases of human infection with *N. salmincola* were reported in 1987, Fritsche et al. reported ten additional cases of human nanophyetiasis. Five presented with gastrointestinal complaints and the other five had unexplained peripheral eosinophilia. Nine out of ten recalled eating inadequately cooked fish. This time, praziquantel was the effective treatment of choice.

In 1990, the first case of human infection with *N. salmincola* without ingestion of raw or undercooked contaminated fish was reported. A man was infected through hand contamination while handling highly infected, fresh-killed, coho salmon. A diagnosis of nanophyetiasis was made based on gastrointestinal discomfort, peripheral blood esoinophilia and a positive stool sample. Treatment with praziquantel proved to be effective again.

None of the human cases of infection with either the North American or Siberian subspecies reveal infection by the *Neorickettsia helminthoeca* carried within the trematode, which was discovered in 1950. Infection by rickettsia helps to explain the more fatal outcome afflicting canids.

Clinical Presentation in Humans

Upon infection with *N. salmincola*, humans are normally asymptomatic. If symptoms are present, they are usually non-specific and mistaken for indication of other gastrointestinal problems.

Symptoms include "diarrhea, unexplained peripheral blood eosinophilia, abdominal discomfort, nausea and vomiting, weight loss, and fatigue." Eggs of *N. salmincola* appear in stools approximately one week after ingestion of infected fish.

Pathology in Dogs

Nanophysiasis in dogs is much more serious than in humans. Scientists noticed almost 200 years ago that dogs that consumed raw fish sometimes died rather quickly. This "salmon poisoning", while associated with the trematode *Nanophyetus salmincola* is not caused by the worm. The sickness is caused by *Neorickettsia helminthoeca,* a rickettsial bacteria that uses the *N. salmincola* as a host. Although only canines are susceptible to the disease raccoons show a raised temperature and lymphatic infection after being infected by the rickettsia, but both soon subside. The incubation period in dogs is 5–7 days, although it may take as long as 33 days. After onset, there is a sharp fever coupled with anorexia, vomiting and dysentery. The rickettsia attacks the canine's lymph system causing enlarging and eventually hemorrhaging many of the lymphnodes. The disease can spread to other tissues such as leucocytes. Death occurs 10–14 days after signs first appear.

Transmission

Nanophyetus salmincola is transmitted most commonly by the ingestion of raw, undercooked, or smoked salmon or steelhead trout. Usually this is meant to be ingestion of the muscle of the fish but there have been cases reported in which the suspected agent of transmission was Steelhead roe. Researchers hypothesize, in fish with especially high worm burdens, that the N. salmincola may migrate to many of the fishes tissues, not just the muscle tissue. In a case in 1990 Nanophyetiasis was diagnosed in an individual that is thought to have acquired the disease by simple handling of fresh-killed salmon. The infected individual, ironically, was a researcher studying N.salmincola in juvenile Coho salmon, had inadvertently initiated the infection by hand-to-mouth contact during the 3 month long study.

Reservoir

The reservoirs for *N. salmincola* are raccoons, mink, and skunks. Reservoirs are organisms that harbor parasites within themselves without suffering any signs of pathology, and spreading the parasites through their natural behavior. For example, raccoons naturally spread *N. salmincola* because they frequently eat fish and defecate parasitic eggs in or near the water, where subsequent larval stages can continue their life cycle.

Vector

Vectors are organisms that transmit parasites from one host to another. *Oxytrema silicula* stream snails are biological vectors for a larval stage of *N. salmincola*. Salmonid and some non-salmonid fish are vectors of the metacercariae of *N. salmincola*. Both fresh and ocean water fish can be parasitic vectors. Fish that act as second intermediate hosts are different species of the families Salmonidae, Cottidae, and Cyprinidae. Among the thirty-four natural and experimental secondary hosts found in scientific literature are the coastal cutthroat trout, rainbow trout, coho salmon, chum salmon, and kokanee salmon. More infection occurs in salmonid fish, rather than non-salmonid fish. In particular, salmonid fish of the genera Salmo, Oncorhynchus, and Salvelinus play a significant role in the *N. salmincola life cycle*. The parasite itself is a vector for *Neorickettsia helminthoeca*.

Definitive Hosts

Definitive hosts include fish-eating birds and mammals. The most common definitive hosts are the domestic dog, cat, and red fox. Humans are also definitive hosts for *N. salmincola*. A long list of experimental definitive hosts include the hamster and wood rat. Interestingly, an experimental study failed to infect two white rats and two white mice. Trematodes exist along the whole length of the small intestine in smaller animals like hamsters, while they exist only in the upper end of the small intestine in larger animals like dogs.

Incubation Period

After ingestion of fish infected with *N. salmincola*, it takes about 1 week for symptoms to occur, namely for eggs to be detected in the stool.

Morphology

Eggs of *N. salmincola* are light brown, ovoid, and operculate at one end, with a small blunt projection at the other end. They measure 0.087 mm to 0.097 mm by 0.038 mm to 0.055 mm. There are normally 5 to 16 eggs in the uterus, and their heaviness allows them to sink rapidly in water.

N. salmincola is a digenic trematode, which means that it is an unsegmented worm that is flattened dorsoventrally. Adult worms alternate shape from "a sphere to a long blunt rod." The worms are

0.8 to 1.1mm long and 0.3 to 0.5 mm wide and are hermaphroditic, having both male and female reproductive organs in the same organism. The two large oval testes are 0.2 to 0.3 mm long and the round ovary is 0.07 to 0.11 mm in diameter. *N. salmincola* has a prominent cirrus pouch, or hollow organ surrounding the male copulatory organ, but no seminal vesicle. True to its character as a trematode, it has an oral sucker 0.15 to 0.18 mm in diameter, and a ventral sucker 0.12 to 0.13 mm in diameter. The oral and ventral suckers are used to grasp and crawl actively about the intestinal tissue of its host, though the worm leaves no extensive mechanical damage.

Life Cycle

The adult lays eggs within the vertebrate host. The vertebrate passes out the eggs in its feces. The first larval stage, the miracidia, develop within the eggs, hatch, and swim away. The miracidia then penetrate the first intermediate host, the *Oxytrema silicula* stream snail. After further development in the stream snail, *N. salmincola* larva develop into rediae, which give rise to cercariae. The cercariae emerge from the snail and penetrate the second intermediate host, the salmonid (some non-salmonid) fish. The parasites develop into metacercaria and encyst within the kidneys, muscles, and fins of the salmonid fish. The parasites enter its final host, including canids and humans, upon ingestion of the infected fish, and develop into adult worms that produce eggs to be passed in the host's feces.

Detailed Information Regarding the Life Cycle Stages

Eggs and Miracidia: The eggs passed in the feces are unembryonated. Experimental studies demonstrate that eggs collected in room water temperature require 75 days to 200 days to hatch., The hatching rate of miracidia from eggs increases with decreasing temperatures, and egg mortality increases with increasing temperatures. Fully developed miracidia within the eggs contract and elongate repeatedly, and newly emerged miracidia swim in "characteristic, long graceful curves." Interestingly, the miracidia seem to have no attraction to host snails, bumping into the snails without attempting to penetrate and infect them.

Rediae: The rediae are the second larval stage of the trematode life cycle, that develops from the miracidum and contains germ cells that develop into cercariae. The rediae are found in the second intermediate host, the snail. Rediae can range from 0.45 mm to 3 mm, and the larger rediae can contain up to 76 cercariae. Rediae and cercariae are found in all tissues of the host snail, but primarily in the gonads and the digestive gland., Rediae destroy the gonads, invade the hepatopancreas, damage it by 1) increased pressure from rapid growth, 2) active ingestion by the parasites, and 3) the disposition of parasitic wastes. Furthermore, parasites take up glycogen and lipids from the hepatopancreas.

Cercariae: The cercariae measure 0.31 mm to 0.47 mm by 0.03 mm to 0.15 mm and live up to 48 hours in water. They have a tendency to infect snails that are at least 2.5 cm in length, though smaller snails have also been observed to shed cercariae. Cercariae shed intermittently by the thousands, entering the mantle cavity of the snail, and drifting out with the "exhalent water current on the right side of the snail's head.", Cercariae from snails in brackish water of a low salinity were found to survive longer than snails in freshwater.20

Once cercariae are shed from the snail, it contracts repeatedly until it contacts a fish and penetrates under its skin within 30 seconds to 2 minutes. The cercariae penetrate further into the renal portal blood system, into the kidney and deeper tissues into the base of the tail. Penetrations sights are easily visible, as the skin, fins, and tails of the fish appear to be heavily eroded and damaged. Cercariae can also indirectly infect the fish, if the fish eat the cercariae orally.

Metacercariae: The cercariae lose their tails in the act of penetration and encyst as metacercariae in almost any tissue of the salmonid fish. The new cyst wall is thin, transparent, and easily ruptured. If the cyst wall breaks, the metacercariae crawl out and re-encyst a few hours later in a tougher, larger cyst wall. While cysts can be found in all tissues of the fish, most encystment occurs in the kidneys and body muscles of the salmonid fish, and in the gills and fins of the non-salmonid fish. Cercariae penetrate less deeply in non-salmonid fish than in salmonid fish. Infected fish experience a decrease in their swimming activity and loss of equilibrium, and it is not uncommon for fish to have as many as 1000 to 2000 metacercariae in its tissues.,

Importantly, metacercariae can be destroyed either by cooking or freezing infected fish.

Snail: The *Oxytrema silicula* host snail is prevalent in coast streams and prefers large rocks, bridges, old planks, and debris on stream bed bottoms. It rarely migrates into shallow water. The infection of snails is high in comparison to the number of cercariae it sheds, since larval development continues slowly over a long period of time.

Evidence of mixed infection varied between studies, but snails with large numbers of N. salmincola were not parasitized by other trematodes. It was also found that monthly incidences of infection in snails ranged from 9-52% after examining over 3000 snails every month for 10 months, and that mature cercariae infected snails in a seasonal manner. Mature cercariae were more likely to infect snails in late April to November.

Neorickettsia Helmintheoca Neorickettsia Helmintheoca: is the etiological agent for salmon poisoning disease, found to be present in all stages of the trematode. It is 0.3 micrometers in size and a purple Giemsa stain indicates that it is Gram negative. Thus far, only canids are susceptible to disease by rickettsia and it is still uncertain how the rickettsia leave the trematode vector and reaches the host tissues. Experiments do show that the bacteria lead to necrosis of lymph follicles, ulceration, and severe hemorrhage in its host.

Diagnostic Tests

1. History of eating raw fish
2. Examination of feces for eggs of *N. salmincola*

Because only a few eggs are contained within each adult worm, patients with light infections are likely to have negative stool tests. Using trichrome stained preparations rather than formalin-ethyl acetate concentrates was more sensitive to identify cases.

Management and Therapy

Praziquantel, 20 mg/kg body weight, three times a day. Praziquantel causes immobilized contraction of the worm, such that it can no longer grasp the intestinal walls, and can be eliminated from the body Three 2-g doses of niclosamide or two 50 mg/kg doses of bithionol have also proven to be effective when Praziquantel was either not available or treatment was refused. However, a single 2-g dose of niclosamine proved to be ineffective treatment therapy, as did 100 mg orally of mebendazole twice a day for three days. If diarrhea recurs, general supplements must also be provided in order to maintain electrolyte balance and meet nutritional requirements.

Epidemiology

Nanophyetus salmincola is limited to the geographic range of its intermediate hosts, primarily the US Pacific Northwest. Stream snails are found west of the Cascade Mountains in Oregon, north to the Olympic Peninsula in Washington, and in part of northern California. It is "the most common systemic trematode in the United States."

Public Health and Prevention Strategies

- Cook fish thoroughly
- Freeze fish for at least 24 hours
- Fish away from known snail endemic places (otherwise make especially sure to cook fish thoroughly)
- Take precautions when handling fish (prevent hand-to-mouth transmission of the metacercariae and be wary of hand contamination from heavily infected fish)
- Check fish periodically for signs of infection, such as cysts or sites of irritation from penetration of the cercarariae (pertinent to fishing companies, fish markets, and restaurants)
- Report sources of infected fish (bodies of water, fish companies, restaurants)
- Eat raw fish only from trusted sources, such as reputable restaurants
- Use molluscicides if can be practically used in smaller bodies of water
- Keep dogs away from streams, making sure they defecate away from snail habitats

Salmon Louse

The salmon louse, *Lepeophtheirus salmonis*, is a species of copepod in the genus *Lepeophtheirus*. It is a sea louse, a parasite living on salmon. It lives off the mucus, skin and blood of the fish.

There has been some research on the problems caused by this species in aquaculture, but little is known about the salmon louse's life in nature. It has been shown, however, that salmon louse infections in fish farming facilities can cause epizootics in wild fish.

Life Cycle

The life cycle consists of 10 stages, with ecdysis in between.

The first two stages are free swimming nauplius I and II, where it has a length between 0.54 and 0.85 mm. The third stage is the copepodit stage, in which the length is ca. 0.7 mm, and the salmon louse attaches itself to the fish.

Stages IV to VII are the chalimus stages. The salmon louse eats from the fish, and grows to a length of 5 mm for the males, 10 mm for the females. Each generation takes about six weeks at a temperature of 10–12 °C (50–54 °F).

In the pre-adult and adult stages (stage VIII to X), the sea louse is now mobile, and it becomes possible to differentiate males and females. The thorax is broad and shield shaped. The abdomen is narrower, and in the females, filled with eggs. The females also have two long egg strings attached to the abdomen. The salmon louse uses its feet to move around on the host or to swim from one host to another.

Saprolegnia

Saprolegnia is a genus of freshwater mould often called a "cotton mould" because of the characteristic white or grey fibrous patches it forms. Current taxonomy puts Saprolegnia as a genus of the heterokonts in the order Saprolegniales.

Habits

Saprolegnia, like most water moulds, is both a saprotroph and necrotroph. Typically feeding on waste from fish or other dead cells, they will also take advantage of creatures that have been injured or compromised eggs. When they inhabit a live animal, they exhibit as a fungal infection known as mycoses.

Saprolegnia is tolerant to a wide range of temperature, 3°C to 33°C, but is more prevalent in lower temperatures. While it is found most frequently in freshwater, it will also tolerate brackish water and even moist soil.

Saprolegnia filaments (hyphae) are long with rounded ends, containing the zoospores. Saprolegnia generally travels in colonies consisting of one or more species. They first form a mass of individual hyphae. When the mass of hyphae grows large enough in size to be seen without use of a microscope, it can be called a mycelium. Colonies are generally white in color, though they may turn grey under the precesence of bacteria or other debris which has become caught in the fibrous mass.

Reproduction

It has a diploid life cycle which includes both sexual and asexual reproduction. In the asexual phase, a spore of *Saprolegnia* releases zoospores. Within a few minutes, this zoospore will encyst, germinate and release another zoospore. This second zoospore has a longer cycle during which most dispersal happens; it will continue to encyst and release a new spore in a process called polyplanetism until it finds a suitable substrate. When a suitable medium is located, the hairs surrounding the spore will lock onto the substrate so that the sexual reproduction phase can start. It is also during this stage of polyplanetism that the *Saprolegnia* are capable of causing infection; the most pathogenic species have tiny hooks at the end of their hairs to enhance their infectious ability.

Once firmly attached, sexual reproduction begins with the production of male and female gametangium, antheridia and oogonium respectively. These unite and fuse together via fertilization tubes. The zygote produced is named an oospore.

Characteristics of Infection

Saprolegnia is generally a secondary pathogen, though in the right circumstances, it can act as primary. It most frequently targets fish, both in the wild and in tank environments. Through cellular necrosis and other epidermal damage, *Saprolegnia* will spread across the surface of its host as a cotton-like film. Though it often stays in the epidermal layers, the mould does not appear to be tissue specific. A *Saprolegnia* infection is usually fatal, eventually causing haemodilution, though the time to death varies depending on the initial site of the infection, rate of growth and the ability of the organism to withstand the stress of the infection.

The extensive mortalities of salmon and migratory trout in the rivers of western Europe in the 1970's and 1980's in the UDN outbreak were probably almost all ultimately caused by the secondary *Saprolegnia* infections.

Historial evidence suggest the *Saprolegnia* species affecting Australian freshwater fish may be an introduced strain, imported in the 1800s with exotic fish species.

Schistocephalus Solidus

Schistocephalus solidus is a tapeworm of fish, fish-eating birds and rodents.

This hermaphroditic parasite belongs to the Eucestoda subclass, of class Cestoda. It parasitizes fish and fish-eating water birds. The fish-eating water bird is the definitive host and reproduction occurs in the bird's intestine. Eggs of the tapeworm are passed with the bird's feces and hatch in the water where the first larval stage, the coracidium, is produced. The coracidium is then ingested by the first intermediate host, a cyclopoid copepod (e.g. *Macrocyclops albidus*). The second larval stage then subsequently develops in the tissue of this host. Within one to two weeks, the infected copepod is ingested by the second intermediate host, the three-spined stickleback, *Gasterosteus aculeatus*. The third larval stage, the plerocercoid, grows in the abdomen of the fish. When the fish is eaten by a bird, the larvae mature and adults start to produce eggs within two days. Reproduction takes place within one to two weeks, after which the parasite dies.

Reproduction for this parasite is unique because of the hermaphroditic nature of this cestode. When it comes to mating, *S. solidus* has three options: (1) self-fertilization (2) breeding with a sibling (3) breeding with an unrelated individual. There are advantages and disadvantages to each of these three options. For example, self-fertilization is advantageous when no mating partners are around but it is disadvantageous because of inbreeding depression—the reduced fitness of offsprings because of the unmasking of deleterious recessive alleles due to the breeding of closely related individuals. Another disadvantage of self-fertilization is not having the ability to exchange genes with other cestodes which leads to increased genetic variation.

Similarly, breeding with a sibling, also known as incestuous mating, also shares some of the same disadvantages as self-fertilization does—inbreeding depression and lack of genetic variation. But incestuous mating is advantageous because it helps maintain gene complexes within the family which may be important for local adaptation purposes. Breeding with unrelated individuals might seem to be most advantageous choice of mating because it increases genetic variation and avoids inbreeding depression, but it could be a very time consuming process.

Data from Schjørring and Jäger's article shows that incestuous mating among S. Solidus significantly reduced the success of egg hatching. Hatching rate of eggs produced by pairs of cestodes that were not related was 3.5 times higher than those that were produced by the sibling pairs. However, those eggs hatched 2.3 times more successfully than eggs that were produced through selfing. The results observed in this experiment corroborates previous studies of S. Solidus showing that eggs produced through selfing correlates with a reduction in hatching success. Although these earlier findings confirmed this study's result, there was a slight disconnect in the rates of egg hatching. Both sources, Christen et al. and Schjørring, found that the hatching rate of egg from unrelated cestodes was 4 times higher than eggs from selfing, while this experiment found that the rate was 8 times higher. This discrepancy can be attributed to study source and the environment in which the experiment was conducted. The previous studies used wild cestodes while this study observed offsprings of wild cestodes bred in the laboratory. It has been speculated that a laboratory environment lowers selective pressures that would otherwise be present in the wild, thus allowing for genes that would be suppressed in the wild to be passed on to the next generation.

Considering the evolutionary disadvantages that may arise from incestuous mating, scientists would suspect cestodes to sexually prefer unrelated mates. Galvani speculates that because S. Solidus are parasites, that the cestodes would be under extreme selective pressures to uphold an evolutionary advantage over the host thus would avoid the genetically unfavorable consequences that would result from incestuous mating.

However, this study contradicts Galvani's prediction showing that even when presented with alternatives, the cestodes were observed to spend a considerable amount of time around the related cestode as can be seen in the plot of Mean Position by Eggs per Mg Cestodes. The data supports the hypothesis that S. Solidus have a preference for siblings and there exists a positive relationship between number of eggs produced and the amount of time spent in close proximity to the related cestode. This study found that pairs that exhibited strong sibling preference produced 12% more eggs in comparison to eggs produced by unrelated pairs.

Kokko and Ots suggests that, despite inbreeding depression, there exists a significant advantage to incestuous mating choices. Their paper states that there are indirect and direct fitness benefits that outweigh the cost of inbreeding. In species where there is low parental investment and sexual encounters are rare and sequential, incestuous breeding is indirectly beneficial. If the prospective mates are related there is an increase mutual interest in finding a resolution with respect to playing the unpreferred sexual role. With less time allotted to conflicting over sexual roles and dominating one another, procreation is more cost-effective. Under these conditions, the greater effectiveness of inbreeding prevails over the detriment of incestuous mating and evolutionarily select for a preference for related mates.

Sea Louse

The sea louse (plural sea lice) is a copepod within the order Siphonostomatoida, family Caligidae. There are 36 genera within this family which include approximately 42 *Lepeophtheirus* and 300 *Caligus* species. Sea lice are marine ectoparasites (external parasites) that feed on the mucus, epidermal tissue, and blood of host marine fish.

This article focuses on the genera *Lepeophtheirus* and *Caligus* which parasitize marine fish, in particular those species that have been recorded on farmed salmon. *Lepeophtheirus salmonis* and various *Caligus* species are adapted to saltwater and are major ectoparasites of farmed and wild Atlantic salmon. Several antiparasitic drugs have been developed for control purposes. Since *L. salmonis* is the major sea louse of concern and has the most known about its biology and interactions with its salmon host, this review will focus on this species.

Caligus rogercresseyi has become a major parasite of concern on salmon farms in Chile, and studies are under way to gain a better understanding of the parasite and the host-parasite interactions. Recent evidence is also emerging that *L. salmonis* in the Atlantic has sufficient genetic differences from *L. salmonis* from the Pacific, suggesting that Atlantic and Pacific *L. salmonis* may have independently co-evolved with Atlantic and Pacific salmonids, respectively.

Diversity and Hosts: Sea Lice on Wild Fish

Most of our understanding of the biology of sea lice, other than the early morphological studies, is based on laboratory studies designed to understand issues associated with sea lice infecting fish on salmon farms. Information on sea lice biology and interactions with wild fish is unfortunately sparse in most areas with a long-term history of open net-cage development, since understanding background levels of sea lice and transfer mechanisms have rarely been a condition of tenure license for farm operators.

Many sea louse species are specific with regards to host genera, for example *L. salmonis* which has high specificity for salmonids, including the widely farmed Atlantic salmon (*Salmo salar*). *Lepeophtheirus salmonis* can parasitize other salmonids to varying degrees, including brown trout (sea trout: *Salmo trutta*), arctic char (*Salvelinus alpinus*), and all species of Pacific salmon. In the case of Pacific salmon, coho, chum, and pink salmon (*O. kisutch*, *O. keta*, and *O. gorbuscha*, respectively) mount strong tissue responses to attaching *L. salmonis*, which lead to rejection within the first week of infection. Pacific *L. salmonis* can also develop, but not complete, its full life cycle on the three-spined stickleback (*Gasterosteus aculeatus L.*). This has not been observed with Atlantic *L. salmonis*.

How planktonic stages of sea lice disperse and find new hosts is still not completely known. Temperature, light and currents are major factors and survival depends on salinity above 25 ‰. It has been hypothesized that *L. salmonis* copepodids migrating upwards towards light and salmon smolt moving downwards at daybreak facilitate in finding a host. Several field and modeling studies have examined copepodid populations in intertidal zones and have shown that planktonic stages can be transported tens of kilometres from their source.

The source of sea lice infections when salmon return from freshwater has always been a mystery. Sea lice die and fall off anadromous fish such as salmonids when they return to freshwater. Atlantic salmon return and travel upstream in the fall to reproduce, while the smolts do not return to saltwater until the next spring.

Pacific salmon return to the marine nearshore starting in June, and finish as late as December, dependent upon species and run timing; whereas the smolts typically outmigrate starting in April, and ending in late August, dependent upon species and run timing. It is possible that sea lice survive on fish that remain in the estuaries or that they transfer to an as yet unknown alternate host to spend the winter. Nonetheless, smolt get infected with sea lice larvae, or even possibly adults, when they enter the estuaries in the spring. It is also not known how sea lice distribute between fish in the wild. Adult stages of *Lepeophtheirus* spp. can transfer under laboratory conditions, but the frequency is low. *Caligus* spp. transfer quite readily and between different species of fish.

Morphology

Lepeophtheirus salmonis tends to be approximately twice the size of most *Caligus* spp. (e.g. *C. elongatus*, *C. clemensi*, etc.). The body consists of 4 regions: cephalothorax, fourth leg-bearing segment, genital complex, and abdomen. The cephalothorax forms a broad shield that includes all of the body segments up to the third-leg bearing segment. It acts like a suction cup in holding the louse on the fish.

All species have mouth parts shaped as a siphon or oral cone (characteristic of the Siphonostomatoida). The second antennae and oral appendages are modified to assist in holding the parasite on the fish. The second antenna is also used by males to grasp the female during copulation. The adult females are always much larger than males and develop a very large genital complex which in many species makes up the majority of the body mass. Two egg strings of 500 to 1000 eggs (*L. salmonis*) that darken with maturation are approximately the same length as the female's body. One female can produce 6-11 pairs of egg strings in a lifetime of approximately 7 months.

Development

Sea lice have both free swimming (planktonic) and parasitic life stages. All stages are separated by moults. The development rate for *L. salmonis* from egg to adult varies from 17 to 72 days depending on temperature. The life cycle of *L. salmonis* is shown in the figure; the sketches of the stages are from Schram.

Eggs hatch into nauplius I which moult to a second naupliar stage; both naupliar stages are non-feeding, depending on yolk reserves for energy, and adapted for swimming. The copepodid stage is the infectious stage and it searches for an appropriate host, likely by chemo- and mechanosensory clues. Currents, salinity, light, and other factors also will assist copepodids in finding a host. Preferred settlement on the fish occurs in areas with the least hydrodynamic disturbance, particularly the fins and other protected areas. Copepodids once attached to a suitable host feed for a period of time prior to moulting to the chalimus I stage. Sea lice continue their development through 3 additional chalimus stages each separated by a moult. A characteristic feature of all 4 chalimus stages is that they are physically attached to the host by a structure referred to as the frontal filament. There are differences in the timing, method of production and the physical structure of the frontal filament between different species of sea lice. With exception of a short period during the moult, the pre-adult and adult stages are mobile on the fish and, in some cases, can move between host fish. Adult females, being larger, occupy relatively flat body surfaces on the posterior ventral and dorsal midlines and may actually out-compete pre-adults and males at these sites.

Feeding Habits

The naupliar and copepodid stages until they locate a host are non-feeding and live on endogenous food stores. Once attached to the host the copepodid stage begins feeding and begins to develop into the first chalimus stage. Copepods and chalimus stages have a developed gastrointestinal tract and feed on host mucus and tissues within range of their attachment. Pre-adult and adult sea lice, especially gravid females, are aggressive feeders, in some cases feeding on blood in addition to tissue and mucus. Blood is often seen in the digestive tract, especially of adult females. *Lepeophtheirus salmonis* are known to secrete large amounts of trypsin into their host's mucus and this may assist in feeding and digestion. Other compounds such as, prostaglandin E2, have also been identified in *L. salmonis* secretions and may assist in feeding and/or serve the parasite in avoiding the immune response of the host by regulating it at the feeding site. It is not known whether sea lice are vectors of disease, but they can be carriers of bacteria and viruses likely obtained from their attachment to and feeding on tissues of contaminated fish.

Disease

Pathology: Sea lice cause physical and enzymatic damage at their sites of attachment and feeding which results in abrasion-like lesions that vary in their nature and severity depending upon a number of factors. These include host species, age and general health of the fish. It is not clear whether stressed fish are particularly prone to infestation. Sea lice infection itself causes a generalized chronic stress response in fish since feeding and attachment cause changes in the mucus consistency and damage the epithelium resulting in loss of blood and fluids, electrolyte changes, and cortisol release. This can decrease salmon immune responses and make them susceptible to other diseases and reduces growth and performance.

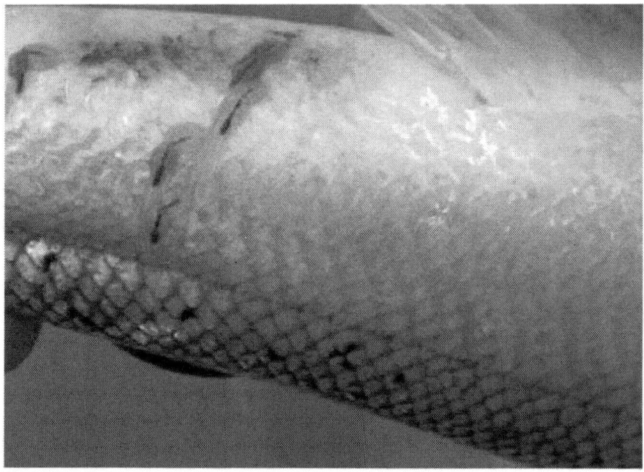

Figure: *Gravid female Lepeophtheirus salmonis on Atlantic salmon, Salmo salar*

The degree of damage is also dependent on the species of sea lice, the developmental stages that are present, and the number of sea lice on a fish. There is little evidence of host tissue responses in Atlantic salmon at the sites of feeding and attachment, regardless of the development stage. In contrast, coho and pink salmon show strong tissue responses to *L. salmonis* characterized by epithelial hyperplasia and inflammation. This results in rejection of the parasite within the first week of infection in these species of salmonids. Heavy infections of farmed Atlantic salmon and wild sockeye salmon (*Oncorhynchus nerka*) by *L. salmonis* can lead to deep lesions, particularly on the head region, even exposing the skull.

Interactions between Wild and Farmed Fish

There is reported concern that sea lice flourishing on salmon farms can spread to nearby wild juvenile salmon and devastate these populations. Sea lice, particularly *Lepeophtheirus salmonis* and various *Caligus* species, including *Caligus clemensi* and *Caligus rogercresseyi*, can cause deadly infestations of both farm-grown and wild salmon. Sea lice migrate and latch onto the skin of wild salmon during free-swimming, planktonic *nauplii* and *copepodid* larval stages, which can persist for several days. Large numbers of highly populated, open-net salmon farms can create exceptionally large concentrations of sea lice; when exposed in river estuaries containing large numbers of open-net farms, many young wild salmon are infected, and do not survive as a result. Adult salmon may survive otherwise critical numbers of sea lice, but small, thin-skinned juvenile salmon migrating to sea are highly vulnerable. On the Pacific coast of Canada, the louse-induced mortality of pink salmon in some regions is commonly over 80%.

Sea trout populations in recent years have also seriously declined due to infestation by sea lice. Several scientific studies have suggested that caged farmed salmon harbour lice to a degree that can destroy surrounding wild salmon populations. Other studies have shown that lice from farmed fish have relatively no effect on wild fish if good husbandry and adequate control measures are carried out (also, see section: Control on salmon farms). Further studies to establish wild-farmed fish interactions are on-going, particularly in Canada, Scotland, Ireland, and Norway. A reference manual with protocol and guidelines for studying wild/cultured fish interactions with sea lice has been published.

Fish Farming

Control on Salmon Farms: This has been reviewed by Pike & Wadsworth, McVicar, and Costello. Integrated pest management programs for sea lice are instituted or recommended in a number of countries, including Canada, Norway, Scotland, and Ireland. Identification of epidemiological factors as potential risk factors for sea lice abundance with effective sea lice monitoring programs have been shown to effectively reduce sea lice levels on salmon farms.

Natural Predators

Cleaner fish, including five species of wrasse (Labridae), are used on fish farms in Norway and to a lesser extent in Scotland, Shetland

and Ireland. Their potential has not been researched in other fish farming regions, such as Pacific and Atlantic Canada or Chile.

Husbandry

Good husbandry techniques include fallowing, removal of dead and sick fish, prevention of net fouling, etc. Bay management plans are in place in most fish farming regions to keep sea lice below a level that could lead to health concerns on the farm or affect wild fish in surrounding waters. These include separation of year classes, counting and recording of sea lice on a prescribed basis, use of parasiticides when sea lice counts increase, and monitoring for resistance to parasiticides.

Salmon Breeding

Early findings suggested genetic variation in the susceptibility of Atlantic salmon to *Caligus elongatus*. Research then began to identify trait markers, and recent studies have shown that susceptibility of Atlantic salmon to *L. salmonis* can be identified to specific families and that there is a link between MHC Class II and susceptibility to lice.

Drugs and Vaccines

The range of therapeutants for farmed fish was limited, often due to regulatory processing limitations. All drugs used have been assessed for environmental impact and risks. The parasiticides are classified into bath and in-feed treatments as follows:

Bath Treatments

There are both advantages and disadvantages to using bath treatments. Bath treatments are more difficult and need more manpower to administer, requiring skirts or tarpaulins to be placed around the cages to contain the drug. Prevention of reinfection is a challenge since it is practically impossible to treat an entire bay in a short time period.

Since the volume of water is imprecise, the required concentration is not guaranteed. Crowding of fish to reduce the volume of drug can also stress the fish.

Recent use of well-boats containing the drugs has reduced both the concentration and environmental concerns, although transferring fish to the well boat and back to the cage can be stressful. The major advantage to bath treatments is that all the fish will be treated equally, in contrast to in-feed treatments where amount of drug ingested can vary due to a number of reasons.

Organophosphates

Organophosphates are acetylcholinesterase inhibitors and cause excitatory paralysis leading to death of sea lice when given as a bath treatment. Dichlorvos was used for many years in Europe and later replaced by azamethiphos, the active ingredient in Salmosan, which is safer for operators to handle. Azamethiphos is water-soluble and broken down relatively quickly in the environment. Resistance to organophosphates began to develop in Norway in the mid 1990s, apparently due to acetylcholinesterases being altered due to mutation. Use has declined considerably with the introduction of SLICE, emamectin benzoate.

Pyrethroids

Pyrethroids are direct stimulators of sodium channels in neuronal cells, inducing rapid depolarization and spastic paralysis leading to death. The effect is specific to the parasite since the drugs used are only slowly absorbed by the host and rapidly metabolized once absorbed. Cypermethrin (Excis, Betamax) and deltamethrin (Alphamax) are the two pyrethroids commonly used to control sea lice. Resistance to pyrethroids has been reported in Norway and appears to be due to a mutation leading to a structural change in the sodium channel which prevents pyrethroids from activating the channel. Use of deltamethrin has been increasing as an alternate treatment with the rise in resistance observed with emamectin benzoate.

Topical Disinfectants

Bathing fish with hydrogen peroxide (350–500 mg/L for 20 min) will remove mobile sea lice from fish. It is environmentally friendly since H_2O_2 dissociates to water and oxygen, but can be toxic to fish, depending on water temperature, as well as to operators. It appears to knock the sea lice off the fish, leaving them capable of reattaching to other fish and reinitiating an infection.

In-feed Treatments

In-feed treatments are easier to administer and pose less environmental risk than bath treatments. Feed is usually coated with the drug and drug distribution to the parasite is dependent on the pharmacokinetics of the drug getting in sufficient quantity to the parasite. The drugs have high selective toxicity for the parasite, are quite lipid soluble so that there is sufficient drug to act for approximately 2 months, and any unmetabolized drug is excreted so slowly that there are little to no environmental concerns.

Avermectins

Avermectins belong to the family of macrocyclic lactones and are the major drugs used as in-feed treatments to kill sea lice. The first avermectin used was ivermectinat doses close to the therapeutic level and was not submitted for legal approval for use on fish by its manufacturer. Ivermectin was toxic to some fish, causing sedation and central nervous system depression due to the drug's ability to cross the blood-brain barrier. Emamectin benzoate, which is the active agent in the formulation SLICE, has been used since 1999 and has a greater safety margin on fish. It is administered at 50 μg/kg/day for 7 days and is effective for two months, killing both chalimus and mobile stages. Withdrawal times vary with jurisdiction from 68 days in Canada to 175 degree days in Norway. Avermectins act by opening glutamate-gated chloride channels in arthropod neuromuscular tissues, causing hyperpolarization and flaccid paralysis leading to death. Resistance has been noted in *Chalimus rogercresseyi* in Chile and *L. salmonis* on North Atlantic fish farms. The resistance is likely due to prolonged use of the drug leading to up-regulation of P-glycoprotein, similar to what has been seen in nematode resistance to macrocyclic lactones.

Growth Regulators

Teflubenzuron, the active agent in the formulation Calicide, is a chitin synthesis inhibitor and prevents moulting. It thus prevents further development of larval stages of sea lice, but has no effect on adults. It has been used only sparingly in sea lice control, largely due to concerns that it may affect the moult cycle of non-target crustaceans, although this has not been shown at the concentrations recommended.

Vaccines

A number of studies are underway to examine various antigens, particularly from the gastrointestinal tract and reproductive endocrine pathways, as vaccine targets, but no vaccine against sea lice has been reported to date.

Other Points of Interest

Branchiurans, family Argulidae, order Arguloida are known as fish lice and parasitize fish in freshwater.

Sphaerothecum Destruens

Sphaerothecum destruens (the rosette agent) is a parasite of fish. It was first discovered in the United States in association with invasive species including topmouth gudgeon but was found to be the causative agent of a disease in the UK affecting salmonid species such as Atlantic salmon and brown trout.

It is thought to pose more of a risk in Europe than in the USA as native species there are more susceptible to the parasite. The disease causes high rates of morbidity and mortality in a number of different salmonid species and can also infect other UK freshwater fish such as bream, carp and roach. The genus *Sphaerothecum* is closely related to the genera *Dermocystidium* and *Rhinosporidium*.

Swim Bladder Disease

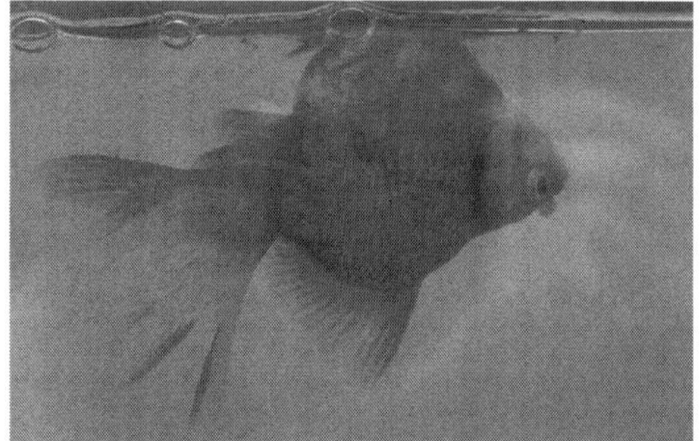

Figure: *Female ryukin goldfish with swim bladder disease*

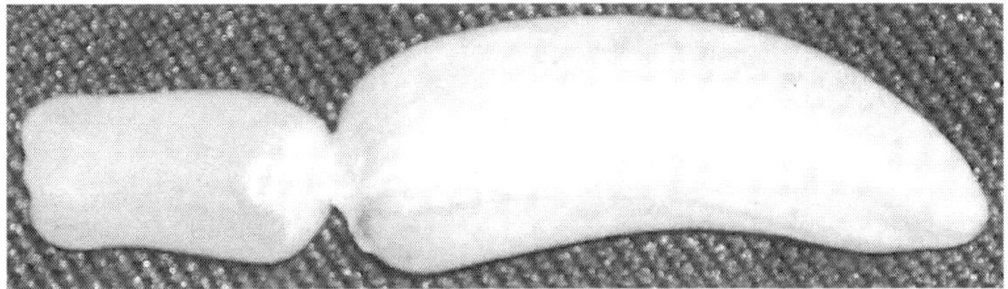

Figure: *The gas bladder of a fish*

Swim bladder disease, also called swim bladder disorder or flipover, is a common ailment in aquarium fish. The swim bladder is an internal gas-filled organ that contributes to the ability of a fish to control its buoyancy, and thus to stay at the current water depth without having to waste energy in swimming. A fish with swim bladder disorder can float nose down tail up, or can float to the top or sink to the bottom of the aquarium.

Causes

Fancy goldfish are among the fish most commonly affected by this disorder. The disease may be caused by intestinal parasites or by constipation induced by high nitrate levels from over feeding.

Remedies

A remedy, which can work within hours, perhaps by countering constipation, is to feed green pea to affected fish. Fish surgeons can also adjust the buoyancy of the fish by placing a stone in the swim bladder or performing a partial removal of the bladder.

Tetracapsuloides Bryosalmonae

Tetracapsuloides bryosalmonae is a myxozoan parasite of salmonid fishes, which causes Proliferative kidney disease (PKD), one of the most serious parasitic diseases of salmonid populations in Europe and North America, which causes losses of up to 90% in infected populations.

Taxonomy

Until the late 1990s, the organism which caused PKD was enigmatic. The "PKX organism", the causative agent of the disease, had been recognized as some form of Malacosporean, but the absence of mature spores in salmonid hosts, the lack of fish to fish transmission, and seasonality of the disease suggest that the life cycle of PKX was completed in another host and that infection of salmonids could be accidental. Korotneff observed a myxozoan in the bryozoan, *Plumatella fungosa*, in 1892, which he described as *Myxosporidium bryozoides*. Myxozoan infection of bryozoans were not reported again until 1996. Ecological investigations of freshwater bryozoans in North America discovered parasitic sacs of a myxozoan species, freely floating in the body cavities of several bryozoans. Molecular analyses indicated that the 18S rDNA sequences of these sacs were indistinguishable from those of PKX, and the PKX organism was scientifically described as *Tetracapsuloides bryosalmonae* Canning, Curry, Feist, Longshaw & Okamura 1999, which has been assigned to a new class, the Malacosporea within the phylum Myxozoa. Around the same time, another group described the PKX organism from Arctic char, *Salvelinus alpinus*, as *Tetracapsuloides renicola* Kent, Khattra, Hedrick & Devlin 2000, but the first given name has priority according to the rules of the binomial nomenclature.

Life Cycle

T. bryosalmonae is highly unusual amongst the myxosporea, in that it uses a bryozoan as an alternate host, rather than an oligochaete or polychaete worm. To date, T. bryosalmonae has been found to parasitize five bryozoan species belonging to the genera Fredericella and Plumatella, all considered to be primitive genera. Problems have occurred in determining the fish host for this species. However, recent work has demonstrated that the parasite cycles between bryozoa and native salmonid species.

Pathology

Proliferative Kidney Disease is characterised by a swollen kidney and spleen, bloody ascites, and pale gills, which indicate the fish is anaemic. Note that these symptoms are common amongst many diseases of fish and do not specifically indicate an infection with *Tetracapsuloides bryosalmonae.*

Distribution

T. bryosalmonae has been recorded in Europe and North America. Phylogenetic analyses of internal transcribed spacer 1 sequences revealed a clade composed of all North American sequences plus a subset of Italian and French sequences. High genetic diversity in North America and the absence of genotypes which are characteristic of the North American clade in the rest of Europe implies that southern Europe was colonized by immigration from North America; however, sequence divergence suggests that this colonization substantially pre-dated human movements of fish. Furthermore, the lack of southern European lineages in the rest of Europe, despite widespread rainbow trout farming, indicates that *T. bryosalmonae* is not transported through fisheries activities. This result contrasts with the commonness of fisheries-related introductions of other pathogens and parasites such as *Myxobolus cerebralis* and *Ceratomyxa shasta.*

Velvet (fish disease)

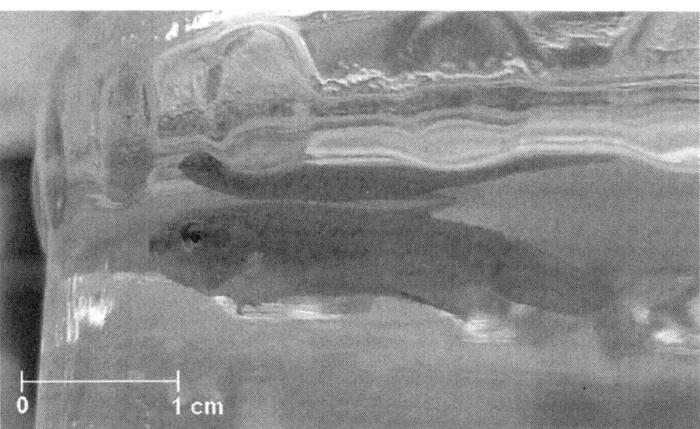

Figure: *Velvet disease*

Velvet disease, also called gold dust disease is a fish disease caused by the dinoflagellate parasites of the genus *Piscinoodinium,* which gives the fish a dusty, slimy look. The disease occurs most commonly in tropical and (to a lesser extent) marine aquaria.

Life Cycle

The single-celled parasite's life cycle can be divided into three major phases. First, as a tomont, the parasite rests at the water's floor and divides into as many as 256 tomites. Second, these juvenile, motile tomites swim about in search of a fish host, meanwhile using photosynthesis to grow, and to fuel their search. Finally, the adolescent tomite finds and enters the slime coat of a host fish, dissolving and consuming the host's cells, and needing only three days to reach full maturity before detaching to become a tomont once more.

Pathology

Velvet (in an aquarium environment) is usually spread by contaminated tanks, fish, and tools (such as nets or testing supplies). There are also rare reports of frozen live foods (such as bloodworms) containing dormant forms of the species. Frequently, however, the parasite is endemic to a fish, and only causes a noticeable "outbreak" after the fish's immune system is compromised for some other reason. The disease is highly contagious and can prove fatal to fish.

Symptoms

Initially, infected fish are known to "flash," or sporadically dart from one end of an aquarium to another, scratching against objects in order to relieve their discomfort. They will also "clamp" their fins very close to their body, and exhibit lethargy. If untreated, a 'dusting' of particles (which are in fact the parasites) will be seen all over the infected fish, ranging in color from brown to gold to green. In the most advanced stages, fish will have difficulty respirating, will often refuse food, and will eventually die of hypoxia due to necrosis of their gill tissue.

Treatment

Sodium chloride is believed to mitigate the reproduction of Velvet, however this treatment is not itself sufficient for the complete eradication of an outbreak. Additional, common medications added directly to the fish's environment include copper sulfate, methylene blue, formalin, malachite green and acriflavin, all of which can be found in common fish medications designed specifically to combat this disease. Additionally, because Velvet parasites derive a portion of their energy from photosynthesis, leaving a tank in total darkness for seven days provides a helpful supplement to chemical curatives. Finally, some enthusiasts recommend raising the water temperature of an infected fish's environment, in order to quicken the life cycle (and subsequent death) of Velvet parasites; however this tactic is not practical for all fish, and may induce immunocompromising stress.

Amnesic Shellfish Poisoning

Amnesic shellfish poisoning (ASP) is a human illness caused by consumption of the marine biotoxin called domoic acid. This toxin is produced naturally by marine diatoms belonging to the genus *Pseudo-nitzschia* and the species *Nitzschia navis-varingica*. When accumulated in high concentrations by shellfish during filter feeding, domoic acid can then be passed on to humans via consumption of the contaminated shellfish. Although human illness due to domoic acid has only been associated with shellfish, the toxin can bioaccumulate in many marine organisms that consume phytoplankton, such as anchovies, and sardines. Intoxication by domoic acid in non-human organisms is frequently referred to as "domoic acid poisoning" or "DAP". In mammals, including humans, domoic acid acts as a neurotoxin, causing permanent short-term memory loss, brain damage, and death in severe cases.

Symptoms and Treatment

In the brain, domoic acid especially damages the hippocampus and amygdaloid nucleus. It damages the neurons by activating AMPA and kainate receptors, causing an influx of calcium. Although calcium flowing into cells is a normal event, the uncontrolled increase of calcium causes the cell to degenerate.

Gastrointestinal symptoms can appear 24 hours after ingestion of affected molluscs. They may include vomiting, nausea, diarrhea, abdominal cramps and haemorrhagic gastritis. In more severe cases, neurological symptoms can take several hours or up to three days to develop. These include headache, dizziness, disorientation, vision disturbances, loss of short-term memory, motor weakness, seizures, profuse respiratory secretions, hiccoughs, unstable blood pressure, cardiac arrhythmia and coma.

People poisoned with very high doses of the toxin or displaying risk factors such as old age and renal failure can die. Death has occurred in 4 of 107 confirmed cases. In a few cases, permanent sequelae included short-term memory loss and peripheral polyneuropathy.

There is no known antidote available for domoic acid, so if symptoms fit the description, it is advised to go quickly to the hospital. Cooking or freezing affected fish or shellfish tissue does not lessen the toxicity.

Discovery

ASP was first discovered in humans late in 1987, when a serious outbreak of food poisoning occurred in eastern Canada. Three elderly patients died and other victims suffered long-term neurological problems. Because the victims suffered from memory loss, the term "amnesic" shellfish poisoning is used. The story made front-page newspaper headlines.

Epidemiologists from Health Canada quickly linked the illnesses to restaurant meals of cultured mussels harvested from one area in Prince Edward Island, a place never before affected by toxic algae. Mouse bioassays on aqueous extracts of the suspect mussels caused death with some unusual neurotoxic symptoms very different from those of paralytic shellfish poisoning toxins and other known toxins. On December 12, 1987, a team of scientists was assembled at the National Research Council of Canada laboratory in Halifax, Nova Scotia. Integrating bioassay-directed fractionation with chemical analysis, the team identified the toxin on the afternoon of December 16, just 4 days after the start of the concerted investigation .

Possible Animal Effects

On June 22, 2006, a California brown pelican, possibly under the influence of domoic acid , flew through the windshield of a car on the Pacific Coast Highway. The phycotoxin is found in the local coastal waters.

Since March 2007, marine mammal and seabird strandings and deaths off the Southern California coast have increased markedly. These incidents have been linked to the recent and dramatic increase of a naturally occurring toxin produced by algae. Most of the animals found dead tested positive for domoic acid.

According to the Channel Islands Marine and Wildlife Institute (CIMWI) , "It is generally accepted that the incidence of problems associated with toxic algae is increasing. Possible reasons to explain this increase include natural mechanisms of species dispersal (currents and tides) to a host of human-related phenomena such as nutrient enrichment (agricultural run-off), climate shifts or transport of algae species via ship ballast water."

In Popular Culture

In the "Bad Fish" episode of *Get a Life* (original air-date: February 2, 1992), Sharon and Gus get amnesia after eating bad shellfish, and Chris seizes the opportunity to convince them that they are his best friends. Domoic acid poisoning may have caused an August 18, 1961 invasion of thousands of frantic seabirds in Capitola and Santa Cruz, California. Director Alfred Hitchcock heard about this invasion while working on his adaptation of Daphne du Maurier novella "The Birds" for his feature film *The Birds* (1963), and asked the *Santa Cruz Sentinel* newspaper for any further news copy as "research for his new thriller".

Diarrhetic Shellfish Poisoning

Diarrhetic shellfish poisoning (DSP) is one of the four recognized symptom types of shellfish poisoning, the others being paralytic shellfish poisoning, neurotoxic shellfish poisoning and amnesic shellfish poisoning. As the name suggests, this syndrome manifests itself as diarrhea, although nausea, vomiting and cramps are all common, too. DSP and its symptoms usually set in within about half an hour of ingesting infected shellfish, and last for about one day. A recent case in France, though, with 20 people consuming oysters manifested itself after 36 hours. The causative poison is okadaic acid, which inhibits intestinal cellular de-phosphorylation. This causes the cells to become very permeable to water and causes profuse diarrhea with a risk of dehydration. As no life-threatening symptoms generally emerge from this, no fatalities from DSP have ever been recorded.

Neurotoxic Shellfish Poisoning

Neurotoxic Shellfish Poisoning (NSP) is caused by the consumption of shellfish contaminated by breve-toxins or brevetoxin analogs. Symptoms in humans include vomiting and nausea and a variety of neurological symptoms such as slurred speech. No fatalities have been reported but there are a number of cases which led to hospitalization.

Paralytic Shellfish Poisoning

Paralytic shellfish poisoning (PSP) is one of the four recognized syndromes of shellfish poisoning, which share some common features and are primarily associated with bivalve mollusks (such as mussels, clams, oysters and scallops). These shellfish are filter feeders and, therefore, accumulate toxins produced by microscopic algae, such as dinoflagellates and diatoms, and cyanobacteria. Human toxicity and mortality can occur after ingestion of these animals, but toxicity is also seen in wild animal populations.

Pathophysiology

The toxins responsible for most shellfish poisonings are water-soluble, heat and acid-stable, and ordinary cooking methods do not eliminate the toxins. The principal toxin responsible for PSP is saxitoxin. Some shellfish can store this toxin for several weeks after a harmful algal bloom passes, but others, such as butter clams, are known to store the toxin for up to two years. Additional toxins are found, such as neosaxiton and gonyautoxins I to IV. All of them act primarily on the nervous system.

PSP can be fatal in extreme cases, particularly in immunocompromised individuals. Children are more susceptible. PSP affects those who come into contact with the affected shellfish by ingestion. Symptoms can appear ten to 30 minutes after ingestion, and include nausea, vomiting, diarrhea, abdominal pain, and tingling or burning lips, gums, tongue, face, neck, arms, legs, and toes. Shortness of breath, dry mouth, a choking feeling, confused or slurred speech, and loss of coordination are also possible.

PSP in Wild Marine Mammals

PSP has been implicated as a possible cause of sea otter mortality and morbidity in Alaska, as one of its primary prey items, the butter clam, (*Saxidonus giganteus*), bioaccumulates saxitoxin as a chemical defense mechanism. . In addition, ingestion of saxitoxin-containing mackerel has been implicated in the death of humpback whales

Additional cases where PSP was suspected as the cause of death in Mediterranean monk seals (*Monachus monachus*) in the Mediterranean Sea have been questioned due to lack of additional testing to rule out other causes of mortality.

Fish Kill

Figure: *There are many causes of fish kill, but oxygen depletion is the most common cause.*

The term fish kill, known also as fish die-off and (in Britain) as fish mortality, is a localized die-off of fish populations which may also be associated with more generalised mortality of aquatic life. The most common cause is reduced oxygen in the water, which in turn may be due to factors such as drought, algae bloom, overpopulation, or a sustained increase in water temperature. Infectious diseases and parasites can also lead to fish kill. Toxicity is a real but far less common cause of fish kill.

Indicators of Environmental Stress

Fish kills are often the first visible signs of environmental stress and are usually investigated as a matter of urgency by environmental agencies to determine the cause of the kill. Many fish species have a relatively low tolerance of variations in environmental conditions and their death is often a potent indicator of problems in their environment that may be affecting other animals and plants and may have a direct impact on other uses of the water such as for drinking water production.

Pollution events may affect fish species and fish age classes in different ways. If it is a cold-related fish kill, juvenile fish and/or species that are not cold-tolerant may be selectively affected. If toxicity is the cause, species are more generally affected and the event may include amphibians and shellfish as well. A reduction in dissolved oxygen may affect larger specimens more than smaller fish as these may be able to access oxygen richer water at the surface, at least for a short time.

Causes

Fish kills may result from a variety of causes. Of known causes, fish kills are most frequently caused by pollution from agricultural runoff or biotoxins. ecological hypoxia (oxygen depletion)is one of the most common natural causes of fish kills. The hypoxic event may be brought on by factors such as algae blooms, droughts, high temperatures and thermal pollution.

Fish kills may also occur due to the presence of disease, agricultural and sewage runoff, oil or hazardous waste spills, sea-quakes, inappropriate re-stocking of fish, poaching with chemicals, underwater explosions, and other catastrophic events that upset a normally stable aquatic population. Because of the difficulty and lack of standard protocol to investigate fish kills, many fish kill cases are designated as having an 'unknown' cause.

Oxygen Depletion

Oxygen enters the water through diffusion. The amount of oxygen that can be dissolved in water depends on the atmospheric pressure, the water temperature and whether the water is salty. For example, at 20°C (68°F) and one atmosphere of pressure, a maximum of 8 mg/l of oxygen can dissolve in sea water (35 mg/l salinity) while a maximum of 9 mg/l of oxygen can dissolve in fresh water. The amount of oxygen that can be dissolved in the water decreases by about 1 mg/l for each 10°C increase in water temperature above 20°C.

Many cold water fish that have evolved to live in clean cold waters become stressed when oxygen concentrations fall below 8 mg/l whilst warm water fish generally need at least 5 ppm (5 mg/l) of dissolved oxygen. Fish can endure short periods of reduced oxygen. Depleted oxygen levels are the most common cause of fish kills. Oxygen levels normally fluctuate even over the course of a day and are affected by weather, temperature, the amount of sunlight available, and the amount of living and dead plant and animal matter in the water. In temperate zones oxygen levels in eutrophic rivers in summertime can exhibit very large diurnal fluctuations with many hours of oxygen supersaturation during daylight followed by oxygen depletion at night. Associated with these photosynthetic rhythms there is a matching pH rhythm as bicarbonate ion is metabolised by plant cells. This can lead to pH stress even when oxygen levels are high.

Additional dissolved organic loads are the most common cause of oxygen depletion and such organic loads may come from sewage, farm waste, tip leachate and many other sources

Diseases and Parasites

Fish are subject to various viruses, bacteria and fungi in addition to parasites such as protozoans, flukes and worms, or crustaceans. These are naturally occurring in many bodies of water, and fish that are stressed for other reasons, such as spawning or suboptimal water quality, are more susceptible. Signs of disease include sores, missing scales or lack of slime, strange growths or visible parasites, and abnormal behaviour – lazy, erratic, gasping at the water surface or floating head, tail or belly up.

For example, since 2004 fish kills have been observed in the Shenandoah River basin in the spring, from the time water temperatures are in the 50s (°F) until they reach the mid-70s. So far, investigators suspect certain bacteria, along with environmental and contaminant factors that may cause immune suppression. In fish farming, where populations are optimized for the available resources, parasites or disease can spread quickly. In channel catfish aquaculture ponds, for example, the "hamburger gill disease" is caused by a protozoan called *Aurantiactinomyxon* and can kill all the fish in an affected pond. In addition to altered behaviour, affected fish have swollen gills that are mottled and have the appearance of ground hamburger meat.

Some early warning signs of fish suffering from disease or parasite infections include:

1. Discolouration, open sores, reddening of the skin, bleeding, black or white spots on the skin
2. Abnormal shape, swollen areas, abnormal lumps, or popeyes
3. Abnormal distribution of the fish such as crowding at the surface, inlet, or pond edges (though crowding at the surface during specific times of day, such as early morning, is more likely a sign of low oxygen)
4. Abnormal activity such as flashing, twisting, whirling, convulsions, loss of buoyancy
5. Listlessness, weakness, sluggishness, lack of activity
6. Loss of appetite or refusal to feed

Toxins

Agricultural runoff, sewage, surface runoff, chemical spills, hazardous waste spills can all potentially lead to water toxicity and fish kill. Some algae species also produce toxins. In Florida, these include *Aphanizomenon*, *Anabaena* and *Microcystis*. Some notable fish kills in Louisiana in the 1950s were due to a specific pesticide called endrin.

Natural instances of toxic conditions can occur, especially in poorly buffered water. Aluminium compound can cause complete fish kills, sometimes associated with autumn turn-over of lakes leading to complex chemical interactions between pH, calcium ions and complex polymeric salts of Aluminium

Human-induced fish kills are unusual, but occasionally a spilled substance causes direct toxicity or a shift in water temperature or pH that can lead to fish kill. For example, in 1997 a phosphate plant in Mulberry, Florida, accidentally dumped 60 million gallons of acidic process water into Skinned Sapling Creek, reducing the pH from about 8 to less than 4 along for about 36 miles and resulting in the death of about 1.3 million fish.

It is often difficult or impossible to determine whether a potential toxin is the direct cause of a fish kill. For example, hundreds of thousands of fish died after an accidental spill of bourbon whiskey into the Kentucky River near Lawrenceburg. However, officials could not determine whether the fish kill was due to the bourbon directly or to oxygen depletion that resulted when aquatic microbes rapidly began to consume and digest the liquor.

Cyanide is a particular toxic compound that has been used to poach fish. In cyanide poisoning the gills turn a distinctive cherry red. Chlorine introduced as alkaline hypochlorite solution is also extremely toxic leaving pale mucilaginous gills and an over-production of mucilage across the whole body. Lime produces similar symptoms but is also often associated with milk eyes.

Algae Blooms and Red Tides

An algae bloom is the appearance of a large amount of algae or scum floating on the surface of a body of water. Algae blooms are a natural occurrence in nutrient-rich lakes and rivers, though sometimes increased nutrient levels leading to algae blooms are due to fertilizer or animal waste runoff. A few species of algae produce toxins, but most fish kills due to algae bloom are a result of decreased oxygen levels. When the algae die, decomposition uses oxygen in the water that would be available to fish. A fish kill in a lake in Estonia in 2002 was attributed to a combination of algae bloom and high temperatures. When people manage algae blooms in fish ponds, it is recommended that treatments be staggered to avoid too much algae dying at once, which may result in a large drop in oxygen content.

Some diseases result in mass die offs. One of the more bizarre and recently discovered diseases produces huge fish kills in shallow marine waters. It is caused by the ambush predator dinoflagellate *Pfiesteria piscicida*. When large numbers of fish, like shoaling forage fish, are in confined situations such as shallow bays, the excretions from the fish encourage this dinoflagellate, which is not normally toxic, to produce free-swimming zoospores. If the fish remain in the area, continuing to provide nourishment, then the zoospores start secreting a neurotoxin. This toxin results in the fish developing bleeding lesions, and their skin flakes off in the water. The dinoflagellates then eat the blood and flakes of tissue while the affected fish die. Fish kills by this dinoflagellate are common, and they may also have been responsible for kills in the past which were thought to have had other causes. Kills like these can be viewed as natural mechanisms for regulating the population of exceptionally abundant fish. The rate at which the kills occur increases as organically polluted land runoff increases. Red tide is the name commonly given to an algal bloom of *Karenia brevis*, a microscopic marine dinoflagellate which is common in Gulf of Mexico waters. In high concentrations it discolours the water which often appears reddish-brown in colour. It produces a toxin which paralyses the central nervous system of fish so they cannot breathe. Dead fish wash up on beaches around Texas and Florida. Humans can also become seriously ill from eating oysters and other shellfish contaminated with the red tide toxin. The term "red tide" is also commonly used to describe harmful algal blooms on the northern east coast of the United States, particularly in the Gulf of Maine. This type of bloom is caused by another species of dinoflagellate known as *Alexandrium fundyense*. These blooms are natural phenomenon, but the exact cause or combination of factors that result in red tides outbreak not fully understood.

Biological Decay

Just as an algae bloom can lead to oxygen depletion, introduction of a large amount of decaying biological material in general to a body of water leads to oxygen depletion as microorganisms use up available oxygen in the process of breaking down organic matter. For example, a 10-mile-long fish kill in September, 2010, in the Sangamon River in Illinois was traced to discharge of animal waste into the river from a large dairy operation. The illegal discharge resulted in a complete kill of fish, frogs, mussels and mudpuppies.

Spawning Fatalities

Figure: *A salmon which has died after spawning*

Some species of fish exhibit mass simultaneous mortality as part of their natural life cycle. Fish kill due to spawning fatalities can occur when fish are exhausted from spawning activities such as courtship, nest building, and the release of eggs or milt (sperm). Fish are generally weaker after spawning and are less resilient than usual to smaller changes in the environment. Examples include the Atlantic salmon and the Sockeye salmon where many of the females routinely die immediately after spawning.

Water Temperature

A fish kill can occur with rapid fluctuations in temperature or sustained high temperatures. Generally, cooler water has the potential to hold more oxygen, so a period of sustained high temperatures can lead to decreased dissolved oxygen in a body of water. An August, 2010, fish kill in Delaware Bay was attributed to low oxygen as a result of high temperatures.

A massive (hundreds of thousands) fish kill at the mouth of the Mississippi River in Louisiana, September, 2010, was attributed to a combination of high temperatures and low tide. Such kills are known to happen in this region in late summer and early fall, but this one was unusually large.

A short period of hot weather can increase temperatures in the surface layer of water, as the warmer water tends to stay near the surface and be further heated by the air. In this case, the top warmer layer may have more oxygen than the lower, cooler layers because it has constant access to atmospheric oxygen. If a heavy wind or cold rain then occurs (usually during the autumn but sometimes in summer), the layers can mix. If the volume of low oxygen water is much greater than the volume in the warm surface layer, this mixing can reduce oxygen levels throughout the water column and lead to fish kill.

Fish kills can also result from a dramatic or prolonged drop in air (and thus, water) temperature. This kind of fish kill is selective - usually the dead fish are species that cannot tolerate cold. This has been observed in cases where a fish native to a more tropical region has been introduced to cooler waters, such as the introduction of the tilapia to bodies of water in Florida. Native to Africa's Nile River, the tilapia stop feeding when water temperatures drop below 60°F and die when it reaches 45°F. Thus, tilapia that have survived and successfully reproduced in Florida are occasionally killed by a winter cold front.

In January, 2011, a selective fish kill affecting an estimated 2 million juvenile spot fish was attributed to a combination of cold stress and overpopulation after a particularly large spawn.

Underwater Explosions

Underwater explosions can lead to fish kill, and fish with swim bladders are more susceptible. Sometimes underwater explosions are used on purpose to induce fish kills, a generally illegal practice known as blast fishing. Underwater explosions may be accidental or planned, such as for construction, seismic testing, mining or blast testing of structures under water. In many places, an assessment of potential effects of underwater explosions on marine life must be completed and preventive measures taken before blasting.

Droughts and Overstocking

Droughts and overstocking can also result in inland fish kills. A drought can lead to lower water volumes so that even if the water contains a high level of dissolved oxygen, the reduced volume may not be enough for the fish population. Droughts often occur in conjunction with high temperatures so that the oxygen carrying capacity of the water may also be reduced. Low river flows also reduce the available dilution for permitted discharges of treated sewage or industrial waste. The reduced dilution increases the organic demand for oxygen further reducing the oxygen concentration available to fish

Overstocking of fish (or an unusually large spawn) can also result in inland fish kills. Fish kill due to insufficient oxygen is really a matter of too much demand and too little supply for whatever reason(s). Recommended stocking densities are available from many sources for bodies of water ranging from a home aquarium or backyard pond to commercial aquaculture facilities.

Estimation

Estimating the magnitude of a kill presents a number of problems.

1. Polluted waters are often very turbid or have low transparency making it difficult or impossible to see fish that have sunk
2. Rivers and streams can move fish downstream out of the investigation area.
3. Small fish and fry can decompose or become buried in sediments very quickly and are lost from the count.
4. Predators and scavengers remove and eat fish.
5. Stressed fish may swim up tributaries and die there
6. Many kills are reported only when dead fish resurface due to decompositional gas formation, often several hours after the kill has occurred.

Some very large fish kills may never be estimated because of these factors. The discharge of red aluminium sludge from a reservoir in Hungary into the Marcai River is acknowledged as causing environmental devastation, The loss of adult fish also can have long term impacts on the success of the fishery as the following year's spawning stock may have been lost and recovery of the pre-kill population may take years to recover. The loss of food supplies or recreational income may be very significant to the local economy.

Prevention and Investigation

Fish kills are difficult to predict. Even when conditions that contribute to fish kill are known to exist, prevention is hard because often conditions cannot be improved and fish cannot be safely removed in time. In small ponds, mechanical aeration and/or removal of decaying matter (such as fallen leaves or dead algae) may be reasonable and effective preventive measures.

Many countries in the developed world have specific provisions in place to encourage the public to report fish kills so that a proper investigation can take place. Investigation of the cause of a kill requires a multi-disciplinary approach including on-site environmental measurements, investigation of inputs, review of meteorology and past history, toxicology, fish autopsy, invertebrate analysis and a robust knowledge of the area and its problems.

Ciguatera

Ciguatera is a foodborne illness caused by eating certain reef fishes whose flesh is contaminated with toxins originally produced by dinoflagellates such as *Gambierdiscus toxicus* which lives in tropical and subtropical waters. These dinoflagellates adhere to coral, algae and seaweed, where they are eaten by herbivorous fish who in turn are eaten by larger carnivorous fish. In this way the toxins move up the food chain and bioaccumulate. *Gambierdiscus toxicus* is the primary dinoflagellate responsible for the production of a number of similar toxins that cause ciguatera.

These toxins include ciguatoxin, maitotoxin, scaritoxin and palytoxin. Predator species near the top of the food chain in tropical and subtropical waters, such as barracudas, snapper, moray eels, parrotfishes, groupers, triggerfishes and amberjacks, are most likely to cause ciguatera poisoning, although many other species cause occasional outbreaks of toxicity. Ciguatoxin is very heat-resistant, so ciguatoxin-laden fish cannot be detoxified by conventional cooking.

Researchers suggest that ciguatera outbreaks caused by cooling climatic conditions propelled the migratory voyages of Polynesians between 1000 and 1400.

Symptoms

Hallmark symptoms of ciguatera in Humans include gastrointestinal and neurological effects. Gastrointestinal symptoms include nausea, vomiting, and diarrhea, usually followed by neurological symptoms such as headaches, muscle aches, paresthesia, numbness, ataxia, and hallucinations. Severe cases of ciguatera can also result in cold allodynia, which is a burning sensation on contact with cold (commonly incorrectly referred to as reversal of hot/cold temperature sensation). Doctors are often at a loss to explain these symptoms and ciguatera poisoning is frequently misdiagnosed as multiple sclerosis.

Dyspareunia and other ciguatera symptoms have developed in otherwise healthy males and females following sexual intercourse with partners suffering ciguatera poisoning, signifying that the toxin may be sexually transmitted. Diarrhea and facial rashes have been reported in breastfed infants of poisoned mothers, it is likely that ciguatera toxins migrate into breast milk.

The symptoms can last from weeks to years, and in extreme cases as long as 20 years, often leading to long-term disability. Most people do recover slowly over time. Often patients recover, but symptoms then reappear. Such relapses can be triggered by consumption of nuts, alcohol, fish or fish-containing products, chicken or eggs, or by exposure to fumes such as those of bleach and other chemicals. Exercise is also a possible trigger. Filipino and Chinese people may possibly be more susceptible.

Detection Methods

Scientific Detection: Currently, multiple laboratory methods are available to detect ciguatoxins, including liquid chromatography-mass spectrometry (LCMS), receptor binding assays (RBA), and neuroblastoma assays (N2A). Although testing is possible, in most cases liquid chromatography-mass spectrometry is insufficient to detect clinically relevant concentrations of ciguatoxin in crude extracts of fish.

Folk Detection

In Northern Australia, where ciguatera is a common problem, two different folk science methods are widely believed to detect whether fish harbor significant ciguatoxin. The first method is that flies will not land on contaminated fish.

The second is that cats display symptoms after eating contaminated fish. A third, less common testing method involves putting a silver coin under the scales of the suspect fish. Only if the coin turns black, is it contaminated. It is not known whether any of these tests produce accurate results.

Folk Remedies

Various Caribbean naturopathic and ritualistic treatments originated in Cuba and nearby islands. The most common old-time remedy involves bed rest subsequent to a guanabana juice enema. Other folk treatments range from directly porting and bleeding the gastrointestinal tract to "cleansing" the diseased with a dove during a Santería ritual. In Puerto Rico, natives drink a tea made from mangrove buttons, purportedly high in B Vitamins, to flush the toxic symptoms from the system. The efficacy of these treatments has never been studied or substantiated.

An account of ciguatera poisoning from a linguistics researcher living on Malakula island, Vanuatu, indicates the local treatment: "We had to go with what local people told us: avoid salt and any seafood. Eat sugary foods. And they gave us a tea made from the roots of ferns growing on tree trunks. I don't know if any of that helped, but after a few weeks, the symptoms faded away."

Treatment

There is no effective treatment or antidote for ciguatera poisoning. The mainstay of treatment is supportive care. There is some evidence that calcium channel blocker type drugs such as Nifedipine and Verapamil are effective in treating some of the symptoms that remain after the initial sickness passes, such as poor circulation and shooting pains through the chest.

These symptoms are due to the cramping of arterial walls caused by maitotoxin Ciguatoxin lowers the threshold for opening voltage-gated sodium channels in synapses of the nervous system. Opening a sodium channel causes depolarization, which could sequentially cause paralysis, heart contraction, and changing the senses of hearing and cold.

Nifedipine is a calcium channel blocker. Some medications such as Amitriptyline may reduce some symptoms, such as fatigue and paresthesia, although benefit does not occur in every case. Steroids and vitamin supplements support the body's recovery rather than directly reducing toxin effects.

Mannitol was once used for poisoning after one study reported symptom reversal. Followup studies in animals and case reports in humans also found benefit from mannitol.

However, a randomized, double-blind clinical trial found no difference between mannitol and normal saline, and based on this result, mannitol is no longer recommended.

Epidemiology

Due to the limited habitats of ciguatoxin-producing microorganisms, ciguatera is common in only subtropical and tropical waters, particularly the Pacific and Caribbean, and usually is associated with fish caught in tropical reef waters. Ciguatoxin is found in over 400 species of reef fish.

Avoiding consumption of all reef fish (any fish living in warm tropical waters) is the only sure way to avoid exposure. Imported fish served in restaurants may contain the toxin and to produce illness which often goes unexplained by physicians unfamiliar with the symptoms of a tropical toxin. Ciguatoxin can also occur in farm-raised salmon.

In 2007, ten people in St. Louis, Missouri developed the disease after eating imported fish.

In February 2008, the U.S. Food and Drug Administration (FDA) traced several outbreaks to the Flower Garden Banks National Marine Sanctuary in the northern Gulf of Mexico, near the Texas–Louisiana shoreline. The FDA advised seafood processors that ciguatera poisoning was "reasonably likely" to occur from eating several species of fish caught as far as 50 miles (80 km) from the sanctuary.

History

Ciguatera was first described by one of the surgeon's mates, William Anderson, on the crew of HMS *Resolution (1771)* (HMS Resolution) in 1774. (NY Times incorrectly gives his name as John)

Disease in Ornamental Fish

Ornamental fish kept in aquariums are susceptible to numerous diseases. The study of fish diseases has remained a rudimentary branch of veterinary medicine. Due to their generally small size and the low cost of replacing diseased or dead fish, the cost of testing and treating diseases is often seen as more trouble than the value of the fish.

Issues in Diagnosis and Treatment

Water change
Food
Nitrates (NO_3^-)
Nitrospira
Nitrites (NO_2^-)
Nitrosomonas
Ammonium (NH_4^+)
Decomposing plant & animal matter

Figure: *Nitrogen cycle*

Due to the artificially limited volume of water and high concentration of fish in most aquarium tanks, communicable diseases often affect most or all fish in a tank. An improper nitrogen cycle, inappropriate aquarium plants and potentially harmful freshwater invertebrates can directly harm or add to the stresses on ornamental fish in a tank. Despite this, many diseases in captive fish can be avoided or prevented through proper water conditions and a well-adjusted ecosystem within the tank.

Etiology

Figure: Henneguya zschokkei *in salmon beard*

Diseases can have a variety of causes, including bacterial infections from an external source such as *Pseudomonas fluorescens* (causing Fin rot and Fish Dropsy), fungal infections (Saprolegnia), Mould infections (Oomycete and Saprolegnia), parasitic disorders (Gyrodactylus salaris, Ichthyophthirius multifiliis, Cryptocaryon, Oodinium causing velvet disease, Brooklynella hostilis, head and lateral line erosion, Glugea, Ceratomyxa shasta, Kudoa thyrsites, Tetracapsuloides bryosalmonae, Ceratomyxa shasta, leeches, nematode, Trematoda, Platyhelminthes and fish louse), viral disorders, metabolic disorders, inappropriate water conditions (insufficient aeration, pH, water hardness, temperature and ammonia poisoning) and malnutrition.

Prevention

Disease cures are almost always more expensive and less effective than simple prevention measures. Often precautions involve maintaining a stable aquarium that is adjusted for the specific species of fish that are kept and not over-crowding a tank or over-feeding the fish. Common preventive strategies include avoiding the introduction of infected fish, invertebrates or plants by quarantining new additions before adding them to an established tank, and discarding water from external sources rather than mixing it with clean water. Similarly, foods for herbivorous fish such as lettuce or cucumbers should be washed before being placed in the tank. Containers that do not have water filters or pumps to circulate water can also increase stress to fish. Other stresses on fish and tanks can include certain chemicals, soaps and detergents, and impacts to tank walls causing shock waves that can damage fish.

Treatment

In some cases the causes of an infection or disease will be obvious (such as fin rot), though in other cases it may be due to water conditions, requiring special testing equipment and chemicals to appropriately adjust the water.

Isolating diseased fish can help prevent the spread of infection to healthy fish in the tank. This also allows the use of chemicals or drugs which may damage the nitrogen cycle, plants or chemical filtration of a properly-functioning tank. Other alternatives include short baths in a bucket that contains the treated water. Salt baths can be used as an antiseptic and fungicide, and will not damaging beneficial bacteria, though ordinary table salt may contain additives which can harm fish. Alternatives include aquarium salt, Kosher salt or rock salt. Gradually raising the temperature of the tank may kill certain parasites, though some diseased fish may be harmed and certain species can not tolerate high temperatures. Aeration is necessary since less oxygen is dissolved in warm water.

Aeromonas Hydrophila

Aeromonas hydrophila is a heterotrophic, Gram-negative, rod shaped bacterium, mainly found in areas with a warm climate. This bacterium can also be found in fresh, salt, marine, estuarine, chlorinated, and un-chlorinated water. *Aeromonas hydrophila* can survive in aerobic and anaerobic environments. This bacterium can digest materials such as gelatin, and hemoglobin. *Aeromonas hydrophila* was isolated from humans and animals in the 1950s. This bacterium is the most well known of the six species of Aeromonas. It is also highly resistant to multiple medications, chlorine, and cold temperatures.

Structure

Aeromonas hydrophila are Gram-negative straight rods with rounded ends (bacilli to coccibacilli shape) usually from .3 to 1 micrometer in width, and 1 to 3 micrometers in length. *Aeromonas hydrophila* does not form endospores, and can grow in temperatures as low as four degrees Celsius. These bacteria are motile by a polar flagella.

Pathology

Because of *Aeromonas hydrophila's* structure, it is very toxic to many organisms. When it enters the body of its victim, it travels through the bloodstream to the first available organ. It produces Aerolysin Cytotoxic Enterotoxin (ACT), a toxin that can cause tissue damage. *Aeromonas hydrophila, Aeromonas caviae,* and *Aeromonas sobria* are all considered to be opportunistic pathogens, meaning they rarely infect healthy individuals. *Aeromonas hydrophila* is widely considered a major fish and amphibian pathogen, and its pathogenicity in humans has been recognised for decades.

Pathogenic Mechanism

It was believed that the pathogenicity of *Aeromonas* spp. is mediated by a number of extracellular proteins such as aerolysin, lipase, chitinase, amylase, gelatinase, hemolysins and enterotoxins. However the pathogenic mechanisms of *Aeromonas* spp. is remain unknown. The recently proposed type III secretion system (TTSS) mediated pathogenic mechanism has been proven to play a pivotal role in *Aeromonas* pathogenesis.

The TTSS is specialized protein secretion machinery that export virulence factors delivered directly to host cells. These factors subvert normal host cell functions in ways that are beneficial to invading bacteria. In contrast to the general secretory pathway, type III secretion system is triggered when a pathogen comes in contact with host cells. ADP-ribosylation toxin is one of the effector molecules secreted by several pathogenic bacteria and translocated through TTSS and delivered into the host cytoplasm leads to interruption of NF-êB pathway, cytoskeletal damage and apoptosis.

This toxin has been characterized in *Aeromonas hydrophila* (human diarrhoeal isolate), *Aeromonas salmonicida* (fish pathogen) and *Aeromonas jandaei* GV17, a pathogenic strain which can cause disease both in human and fish.

Occurrence of Exposure

Aeromonas hydrophila infections occur most during environmental changes, stressors, change in the temperature, in contaminated environments, and when an organism is already infected with a virus or another bacterium. It can also be ingested through food products that have already been infected with the bacterium, such as seafood, meats, and even certain vegetables such as sprouts.

Fish and Amphibians

Aeromonas hydrophila is associated with diseases mainly found in fish and amphibians, because these organisms live in aquatic environments. It is linked to a disease found in frogs called red leg, which causes internal, sometimes fatal hemorrhaging. When infected with Aeromonas hydrophila, fish develop ulcers, tail rot, fin rot, and hemorrhagic septicaemia. Hemorrhagic septicaemia causes lesions that lead to scale shedding, hemorrhages in the gills and anal area, ulcers, exophthalmia, and abdominal swelling.

Human Diseases

Aeromonas hydrophila is not as pathogenic to humans as it is to fish and amphibians. One of the diseases it can cause in humans is gastroenteritis. This disease can affect anyone, but it occurs most in young children and people who have compromised immune systems or growth problems. This bacterium is linked to two types of gastroenteritis. The first type is a disease similar to cholera, which causes rice-water diarrhea. The other type of disease is dysenteric gastroenteritis, which causes loose stools filled with blood and mucus. Dysenteric gastroenteritis is the most severe out of the two types, and can last for multiple weeks. *Aeromonas hydrophila* is also associated with cellulitis, an infection that causes inflammation in the skin tissue. It also causes diseases such as myonecrosis and eczema in people with compromised immune systems.

Outbreaks

Though *Aeromonas hydrophila* can cause serious diseases, there have never been serious outbreaks. There was an outbreak inside the intestinal tract of lizards in Puerto Rico. There were 116 different strains found in the lizards. On May 1, 1988 there was a small *Aeromonas hydrophila* outbreak in California. There were 225 isolates and 219 patients admitted in the hospital because of the bacterium. Confidential Morbidity Report cards were used to report the cases of the bacterium to the local health departments. Investigations were conducted, and reports were sent to the California department of health services for diagnosis and methods in treatment.

Treatments

Aeromonas hydrophila can be eliminated using one percent sodium hypochlorite solution and two percent calcium hypochlorite solution.

Antibiotic agents such as chloramphenicol, florenicol, tetracycline, sulfonamide, nitrofuran derivatives, and pyrodinecarboxylic acids are used to eliminate and control the infection of *Aeromonas hydrophila*. Terramycin is placed in fish food during hatchery operations as another chemotherapeutic agent in preventing *Aeromonas hydrophila*.

Preventing Infection

It is ill-advised to transfer fish from hatchery to hatchery without any sanitation. Hatchery workers should clean the fish, and check for bacterial infection between each operation. To avoid contamination oxygen levels in fish should be maintained, and fish should always be handled gently, to avoid injury. Prophylactic treatments can also be used when trying to prevent *Aermonas hydrophila*. These treatments include disinfectants and Acriflavine.

Bacterial Cold Water Disease

Flavobacterium psychrophilum (previously known as Cytophaga psychrophila) is a member of the Flavobacteriaceae family. It is a gram-negative rod-shaped bacterial pathogen that can cause bacterial cold water disease (BCWD) in salmonid fish. The disease typically occurs at temperatures below 13p C, and disease can be seen in any area with water temperature consistently below 15p C. Although salmon is the most commonly affected species, most fish are susceptible to infection. This disease is not zoonotic. Asymptomatic carrier fish and contaminated water provide reservoirs for disease. Transmission is mainly horizontal, but vertical transmission can also occur.

BCWD may be referred to by a number of other names including cold water disease, peduncle disease, fit rot, tail rot and rainbow trout fry mortality syndrome.

Clinical Signs and Diagnosis

Fish infected with typical BCWD have lesions on the skin and fins. Fins may appear dark, torn, split, ragged, frayed and may even be lost completely. Affected fish are often lethargic and stop feeding. Infection may spread systemically. Salmonid fish can also get a chronic form of BCWD following recovery from typical BCWD. It is characterised by erratic "corkscrew" swimming, blackened tails and spinal deformities.

In Rainbow Trout Fry Syndrome, acute disease with high mortality rates occurs. Infected fish may show signs of lethargy, inappetance and exopthalmos before death.

A presumptive diagnosis can be made based on the history, clinical signs, pattern of mortality and water temperature, especially if there is a history of the disease in the area. The organism can be cultured for definitive diagnosis. Alternatively, histology should show periostitis, osteitis, meningitis and ganglioneuritis.

Treatment and Control

Quaternary ammonium compounds can be to the water of infected adult fish and fry. Alternatively, the antibiotic 'teramycin' can be given to adults, fry and broodstock. To prevent the disease, its is necessary to ensure water is pathogen-free and that water hardening is completed effectively for eggs.

Flavobacteria

The class Flavobacteria is composed of a single order of environmental bacteria. Flavobacteria are a group of commensal bacteria and opportunistic pathogens. *Flavobacterium psychrophilum* causes the septicemic diseases rainbow trout fry syndrome and bacterial cold water disease. According to Bernardet *et al.*, Flavobacteria are gram-negative aerobic rods, 2–5 um long, 0.3–0.5 um wide, with rounded or tapered ends that are motile by gliding, yellow (cream to orange) colonies on agar, decompose several polysaccharides but not cellulose, G+C contents of 32–37 %, and are widely distributed in soil and freshwater habitats. The type species is *F. aquatile.*

Flavobacteriaceae

The family Flavobacteriaceae is composed of environmental bacteria. Most species are aerobic, some are microaerobic to anaerobic, for example *Ornithobacterium*, *Capnocytophaga* and *Coenonia.*

Salmonidae

Salmonidae is a family of ray-finned fish, the only living family currently placed in the order Salmoniformes. It includes salmon, trout, chars, freshwater whitefishes and graylings. The Atlantic salmon and trout of genus *Salmo* give the family and order their names.

Salmonids have a relatively primitive appearance among the teleost fish, with the pelvic fins being placed far back, and an adipose fin towards the rear of the back. They are slender fish, with rounded scales and a forked tail. Their mouths contain a single row of sharp teeth. Although the smallest species is just 13 centimetres (5.1 in) long as an adult, most are much larger, and the largest can reach 2 metres (6.6 ft). All salmonids spawn in fresh water, but in many cases, the fish spend most of their life at sea, returning to the rivers only to reproduce. This type of life cycle is described as anadromous. They are predators, feeding on small crustaceans, aquatic insects, and smaller fish.

Evolution

Current salmonids arose from three lineages: whitefish (Coregoninae), graylings (Thymallinae), and the char, trout and salmons (Salmoninae). Generally, it is accepted that all three lineages share a suite of derived traits indicating a monophyletic group.

Salmonidae first appear in the fossil record in the middle Eocene with the fossil *Eosalmo driftwoodensis* first described from fossils found at Driftwood Creek, central British Columbia. This genus shares traits found in the Salmoninae, whitefish and grayling lineages. Hence, *E. driftwoodensis* is an archaic salmonid, representing an important stage in salmonid evolution.

A gap appears in the salmonine fossil record after *E. driftwoodensis*; until the late Miocene (~7 m.y.a.) trout-like fossils appear in Idaho, in the Clarkia Lake beds. Several of these species appear to be *Oncorhynchus*—the current genus for Pacific salmon and some trout. The presence of these species so far inland established that *Oncorhynchus* was not only present in the Pacific drainages before the beginning of the Pliocene (~5–6 m.y.a.), but also that rainbow and cutthroat trout, and Pacific salmon lineages had diverged before the beginning of the Pliocene. Consequently, the split between *Oncorhynchus* and *Salmo* (Atlantic salmon) must have occurred well before the Pliocene. Suggestions have gone back as far as the early Miocene (~20 m.y.a.).

Classification

Together with the closely related Esociformes (the pikes and related fishes), Osmeriformes (e.g. smelts) and Argentiniformes, the Salmoniformes comprise the superorder Protacanthopterygii.

The Salmonidae (and Salmoniformes) are divided into three subfamilies and around ten genera:

Order Salmoniformes

- Family: Salmonidae
    - Subfamily: Coregoninae
        - *Coregonus* - Whitefishes (70 species)
        - *Prosopium* - round whitefishes (6 species)
        - *Stenodus* - inconnu (1 species)
    - Subfamily: Thymallinae
        - *Thymallus* - Graylings (12 species)
    - Subfamily: Salmoninae
        - *Brachymystax* - lenoks (3 species)
        - *Hucho* (5 species)
        - *Oncorhynchus* - Pacific salmon and trout (14 species)
        - *Salmo* - Atlantic salmon and trout (29 species)
        - *Salvelinus* - Char and trout (e.g. Brook trout, Lake trout) (49 species)
        - *Salvethymus* (1 species)
        - *Acantholingua* (1 species)

Bath treatment (fishkeeping)

Bath treatments are disease treatments that originated since the earliest days of goldfish culture. They are easy to carry out. One of the most effective procedure is called the *salt bath* which is quite effective in eradicating ciliated parasites from the fish.

However, there are also useless, thus not recommended, bath treatments such as the use of certain antibiotics or vitamins in the bath. One disadvantageous limitation of bath treatments is that it harms the beneficial bacteria in the biological filter of the aquarium.

Bath treatment can either be short-term (a "dip") or long-term (from a few hours to 12 or 24 hours or longer).

Medications Used in Bath Treatments

- *Salt:* the most effective bath treatment, and is used to eliminate ciliated protozoan parasites (including ich in small fish); also used to curb the absorption of nitrite, and to reduce the osmotic pressure exerted by fresh-water on any hole in the skin or gill; Dosage and dosing: To produce a 0.3% salt solution (a treatment dosage against most protozoan parasites), 1 tablespoon of salt is used per gallon of water; to produce a 0.6% salt solution (a treatment dosage against Trichodina), 2 tablespoons of salt for every one gallon of water; and to produce a 0.9% salt solution, 3 tablespoons of salt for every one gallon of water
- *Acriflavine:* a powerful dye used for treating protozoans, fungus, lymphocytes, Oodinium and Hexamita
- *Chloramine T:* a quaternary ammonium compound used to control bacterial gill disease and flukes
- *Ionic copper:* free-form copper used to kill bacteria and parasites
- *Dimilin:* an insecticide used against Lernea, Argulus and Ergasilus
- *Droncit:* a pill form of praziquantel used to clear flukes and worms
- *Flagyl:* this is actually metronidazole and is a treatment of choice for Hexamita and Spironucleus
- *Fluke Tabs:* used against fluke infestation
- *Formalin:* this is formaldehyde gas in water used to as a fungicide and treatment against some bacterial infections, ciliated protozoans and flukes
- *Furazone green:* a combination of two or three furan antibiotics
- *Malachite green:* a powerful dye that eliminates parasites
- *Methylene blue:* a basic thiazine dye used to treat fungal infections

- *Pond Health Guard:* a modified version of formalin used to treat flukes
- *Potassium permanganate:* a caustic alkali that is effective in treating flukes, fungal infections, bacterial gill disease, bacterial infections of the body and fins, and ciliated protozoans infestations except ich
- *Program:* is actually lufenuron which is effective against Argulus, Lernea and Ergasilus
- *Tramisole:* is actually levamisole phosphate which is a deworming medication

Aquarium Diseases

The following is a list of aquarium diseases. New fish can sometimes introduce diseases to aquaria, and these can be difficult to diagnose and treat. Most fish diseases are also aggravated when the fish is stressed. Common aquarium diseases include the following:

- *Amylodinium* (marine velvet)
- Anchor worms
- Columnaris
- *Cryptocaryon* (marine ick)
- *Dactylogyrus* (gill flukes)
- Dropsy
- *Mycobacterium marinum* (fish tubercolosis)
- Fin rot
- *Gyrodactylus* (skin flukes)
- Skin or Gill Flukes
- *Ichthyophthirius* (white spot or ick)
- Velvet Disease, including *Oodinium*
- Hexamita (hole in the head)
- Lymphocystis
- *Flexibacter columnaris*
- *Piscine tuberculosis*

The Quarantine

The goal of quarantine is to prevent problems in the main tank due to sickness. A quarantine tank should be used before to introduce any newly acquired animals in the main tank and to treat fish that are already sick. By doing this, the aquarist can avoid the spread of the disease and make it easier to treat the fish.

Brooklynella Hostilis

Brooklynella hostilis is a marine parasite that causes a condition known as "Clownfish disease". It is known to be found in marine aquariums.

Carnivorous Protest

Predatory dinoflagellates are predatory heterotrophic or mixotrophic alveolate protists that derive some or most of their nutrients from digesting other organisms. About one half of dinoflagellates lack photosynthetic pigments and specialize in consuming other eukaryotic cells, and even photosynthetic forms are often predatory.

Organisms that derive their nutrition in this manner include *Oxyrrhis marina*, which feeds phagocytically on phytoplankton, *Polykrikos kofoidii*, which feeds on several species of red-tide and/or toxic dinoflagellates, *Ceratium furca*, which is primarily photosynthetic but also capable of ingesting other protists such as ciliates, *Cochlodinium polykrikoides*, which feeds on phytoplankton, *Gambierdiscus toxicus*, which feeds on algae and produces a toxin that causes ciguatera fish poisoning when ingested, and *Pfiesteria* and related species such as *Luciella masanensis*, which feed on diverse prey including fish skin and human blood cells. Predatory dinoflagellates can kill their prey by releasing toxins or phagocytize small prey directly.

Some predatory algae have evolved extreme survival strategies. For example, *Oxyrrhis marina* can turn cannibalistic on its own species when no suitable non-self prey is available, and *Pfiesteria* and related species have been discovered to kill and feed on fish, and since have been (mistakenly) referred to as carnivorous "algae" by the media.

Usage in the Popular Media

The media has applied the term carnivorous or predatory algae mainly to *Pfiesteria piscicida*, *Pfiesteria shumwayae* and other *Pfiesteria*-like dinoflagellates implicated in harmful algal blooms and fish kills. *Pfiesteria* as an "ambush predator" utilizes a "hit and run" feeding strategy by releasing a toxin that paralyzes the respiratory systems of susceptible fish, such as menhaden, thus causing death by suffocation. It then consumes the tissue sloughed off its dead prey. *Pfiesteria piscicida* (Latin: *fish killer*) has been blamed for killing more than one billion fish in the Neuse and Pamlico river estuaries in North Carolina and causing skin lesions in humans in the 1990s.

It has been described as "skinning fish alive to feed on their flesh" or chemically sensing fish and producing lethal toxins to kill their prey and feed off the decaying remains. Its deadly nature has led to *Pfiesteria* being referred to as "killer algae" and has earned the organism the reputation as the "*T. rex* of the dinoflagellate world" or "the Cell from Hell."

"Pfiesteria Hysteria"

The prominent and exaggerating media coverage of *Pfiesteria* as carnivorous algae attacking fish and humans has been implicated in causing "*Pfiesteria* hysteria" in the Chesapeake Bay in 1997 resulting in an apparent outbreak of human illness in the Pocomoke region in Maryland. However, a study published the following year concluded the symptoms were unlikely to be caused by mass hysteria.

In Popular Culture

During the media coverage in the 1990s, *Pfiesteria* has been referred to as "super villain" and subsequently has been used as such in several fictional works. A *Pfiesteria* subspecies killing humans featured in James Powlik's 1999 environmental thriller *Sea Change*. In Frank Schätzing's 2004 science fiction novel The Swarm, lobsters and crabs spread the killer alga *Pfiesteria homicida* to humans.

In Yann Martel's 2001 novel *Life of Pi*, the protagonist encounters a floating island of carnivorous algae inhabited by meerkats while shipwrecked in the Pacific Ocean. At a book reading in Calgary, Canada, Martel explained that the carnivorous algae island had the purpose of representing the more fantastical of two competing stories in his novel and challenge the reader to a "leap of faith."

In the 2005 National Geographic TV show *Extraterrestrial*, the alien organism termed *Hysteria* combines characteristics of *Pfiesteria* with those of cellular slime molds. Like *Pfiesteria*, *Hysteria* is a unicellular, microscopic predator capable of producing a paralytic toxin. Like cellular slime molds, it can release chemical stress signals that cause the cells to aggregate into a swarm which allows the newly formed superorganism to feed on much larger animals and produce a fruiting body that releases spores for reproduction.

Channel Catfish Virus

Channel Catfish virus (Ictalurid herpesvirus 1) is a member of the *Alloherpesviridae* family that causes disease in catfish. Infection with Channel catfish viral disease (CCVD) can cause significant economic loss in channel catfish farms. The disease is endemic in the USA and there are reports of the virus in Honduras and Russia. Transmission is both horizontal and vertical. Disease occurs in fish less than a year old that weigh less than 10 grams, and the critical environmental factor is water temperature, with disease occurring during warm weather. Channel catfish and blue catfish are the only fish that are susceptible to CCVD and outbreaks are normally seen in farmed fish. Reservoirs of disease are clinically affected fish and recovered covert carriers.

Clinical Signs

Affected fish have pot-bellied appearance, haemorrhages on the fins and musculature and exopthalmos. Affected fish seem off balance and tend to swim erratically or close to the surface, eventually sinking to the bottom.

Fish that survive the infection have lifelong protective immunity but remain latent carriers of CCV. This is a significant source of disease for vulnerable fish.

Increased mortality in young catfish during warm weather, especially after stress is suggestive of CCVD.

Diagnosis

CCV can be detected in water containing infected fish and organs of diseased fish by virus neutralisation, fluorescent antibody testing, ELISA or PCR. FAT and ELISA should be used for diagnosis of clinically infected fish while virus neutralisation or PCR should be used to to detect carrier fish. Lesions are seen on the liver, kidney and many other internal organs both histologically and grossly on postmortem examination.

Treatment and Control

There is no available treatment. Stress and high stocking-densities should be avoided to reduce disease occurrence. Appropriate quarantine and hygiene measures should be employed to prevent spread of disease. The virus is sensitive to acidic pH, heat and UV light and is inactivated by pond mud and sea water.

Cryptobia

Cryptobia species are protozoa that can lead to disease in fish. *C. branchialis* is an ectoparasite that lives on the skin or gills of fish, and the other four pathogenic species are endoparasites that live in the circulatory system or intestines. Marine and freshwater fish can be infected and the disease is most important in salmonid fish. The disease is observed in most continents including the USA, Eastern Europe and Mexico. Bloodfeeding leeches are implicated in the transmission of the disease for those species that reside in the bloodstream.

Clinical Signs and Diagnosis

C. branchialis infects the gills and can lead to skin changes, anorexia and death.

C. iubilans resides in the intestines and leads to granulomatous inflammation in the abdominal organs resulting in weight loss and death.

C. salmositica, C. borreli and *C. bullocki* are all blood parasites that lead to anaemia and lesions in the haematopoietic tissues.

Skin or gill biopsies and blood samples can be examined for moving flagellated organisms.

An antibody response can be detected with ELISA testing and fluorescent antibody testing for certain Cryptobia species.

Treatment and Control

Chemical treatment has been effective using isometamidium chloride. A vaccine against *C. salmositica* is available and lasts up to 2 years. Fish can be selected and bred for resistance as a method of control of the disease.

Edwardsiella Ictaluri

Edwardsiella ictaluri (also known as Enteric Septicaemia of Catfish, Hole in the Head Disease, and ESC) is a member of the *Enterobacteriaceae* family. The bacterium is a short, gram negative, pleomorphic rod with flagella. It causes the disease enteric septicaemia of catfish (ESC), which infects a variety of fish species (including many catfish species, knifefish and barbs). The bacteria can cause either acute septicaemia or chronic encephalitis in infected fish. Outbreaks normally occur in spring and autumn.

E. ictaluri can be found in Asia and the United States, being of particular economic importance in the U.S. It is not a zoonosis.

Clinical Signs and Diagnosis

Acute ESC infection causes an acute septicaemia that presents as multiple petechial haemorrhages that develop into depigmented ulcers. Additional clinical signs include abnormal behavior, exophthalmos, hemorrhagic gastroenteritis, edema and ascites. Chronic ESC infection causes a chronic encephalitis. Clinical signs include abnormal behavior, abnormal swimming patterns, swelling and ulceration of the head and death. Any fish that survive the infection become latent carriers of the disease. A presumptive diagnosis can be made based on the clinical signs alone but PCR, indirect immunofluorescence, bacterial culture and ELISA can be used to definitively diagnose the disease.

Treatment and Control

Several antibiotics can be used to treat the disease, but there are reports of resistance emerging. Vaccination can be used to prevent disease. Management factors such as reducing stocking density and stress should be considered.

Edwardsiella Tarda

Edwardsiella tarda is a member of the Enterobacteriaceae family. The bacterium is a facultatively anaerobic, small, motile, gram negative, straight rod with flagella. Infection causes Edwardsiella septicemia (also known as ES, edwardsiellosis, emphysematous putrefactive disease of catfish, fish gangrene and red disease) in channel fish, eels and flounder. It is a zoonosis and can infect a variety of animals including fish, amphibians, reptiles and mammals. *E.tarda* has a worldwide distribution - it is found in mud and the intestine of fish and other marine animals. It is spread by carrier animal faeces.

Clinical Signs and Diagnosis

Infection can cause organomegaly, ocular disease, rectal prolapse, ecchymosis and erosions on the skin, inflammation of the gills, oedema, ascites, abnormal behavior and haemorrhage throughout the body. On postmortem fish are normally pale with widespread petechial haemorrhage and abscessation. It can cause a variety of signs in humans including gastroenteritis, meningitis and peritonitis. A presumptive diagnosis may be made based on the history, clinical signs and autopsy findings. However *E. tarda* can be cultured on specific growth mediums such as brain–heart infusion agar and techniques such as indirect fluorescent antibody testing, ELISA and loop-mediated isothermal amplification (LAMP) can be used to confirm diagnosis.

Treatment and Control

Antibiotics should be used to treat infected fish. Control of the disease is achieved by vaccination. There are three vaccine types and they should all be administered by water bath. Management factors such as reducing stress and stocking density and maintaining a high water quality can help prevent disease.

External Bacterial Infection (fish)

Symptoms

There are a great deal of possible symptoms associated with this disorder. There may be spots on the body which appear red or orange. Watch for red streaks on the surface on the body. Dropsy (bloating) is also a sign of a bacterial disorder. "False fungal infections" look like fungus but is actually a bacterial infection known as Columnaris. These symptoms may include a white or gray film on the body. Treatment: There are a number of effective treatments for many stains of bacterial infections. Three of the most common are tetracycline, penicillin and naladixic acid. Salt baths are another effective treatment.

Information

Bacterial infections are often difficult to diagnose due to the many different types. Orange or red streaks on the body is usually the only fool-proof method for the determination of a bacterial infection.

Lymphocystis

Lymphocystis is a common viral disease of freshwater and saltwater fish. Aquarists often come across this virus when their fish are stressed such as when put into a new environment and the virus is able to grow. The fish starts growing small white pin-prick like growths on their fins or skin and this is often mistaken for Ich/Ick (Ichthyophthirius multifiliis) in the early stages. It soon clumps together to form a cauliflower-like growth on the skin, fins, and occasional gills.

There is no known treatment for this virus, though some aquarists recommend surgery to remove the affected area if it is very serious. Eventually the growths inhibit the fish's ability to swim, breathe or eat, and secondary bacterial infections usually kills the fish.

Usually the best cure is to simply give the fish a stress free life, a weekly bacteria treatment and the virus will slowly subside and the fins will repair themselves. This can take many months.

6

Fish Diseases and Parasites

Like humans and other animals, fish suffer from diseases and parasites. Fish defences against disease are specific and non-specific. Non-specific defences include skin and scales, as well as the mucus layer secreted by the epidermis that traps microorganisms and inhibits their growth. If pathogens breach these defences, fish can develop inflammatory responses that increase the flow of blood to infected areas and deliver white blood cells that attempt to destroy the pathogens. Specific defences are specialised responses to particular pathogens recognised by the fish's body, that is immune responses. In recent years, vaccines have become widely used in aquaculture and ornamental fish, for example vaccines for furunculosis in farmed salmon and koi herpes virus in koi. Some commercially important fish diseases are VHS, ich and whirling disease.

Disease

All fish carry pathogens and parasites. Usually this is at some cost to the fish. If the cost is sufficiently high, then the impacts can be characterised as a disease. However disease in fish is not understood well. What is known about fish disease often relates to aquaria fish, and more recently, to farmed fish. Disease is a prime agent affecting fish mortality, especially when fish are young. Fish can limit the impacts of pathogens and parasites with behavioural or biochemical means, and such fish have reproductive advantages. Interacting factors result in low grade infection becoming fatal diseases. In particular, things that causes stress, such as natural droughts or pollution or predators, can precipitate outbreak of disease.

Disease can also be particularly problematic when pathogens and parasites carried by introduced species affect native species. An introduced species may find invading easier if potential predators and competitors have been decimated by disease.

Pathogens can cause fish diseases such as:

- viral disorders
- bacterial infections, such as *Pseudomonas fluorescens* leading to fin rot and fish dropsy
- fungal infections, such as saprolegnia
- mould infections, such as oomycete and saprolegnia

Parasites

Parasites in fish are a natural occurrence and common. Parasites can provide information about host population ecology. In fisheries biology, for example, parasite communities can be used to distinguish distinct populations of the same fish species co-inhabiting a region.

Additionally, parasites possess a variety of specialized traits and life-history strategies that enable them to colonize hosts. Understanding these aspects of parasite ecology, of interest in their own right, can illuminate parasite-avoidance strategies employed by hosts.

Usually parasites (and pathogens) need to avoid killing their hosts, since extinct hosts can mean extinct parasites. Evolutionary constraints may operate so parasites avoid killing their hosts, or the natural variability in host defensive strategies may suffice to keep host populations viable.

Parasite infections can impair the courtship dance of male threespine sticklebacks. When that happens, the females reject them, suggesting a strong mechanism for the selection of parasite resistance." However not all parasites want to keep their hosts alive, and there are parasites with multistage life cycles who go to some trouble to kill their host.

For example, some tapeworms make some fish behave in such a way that a predatory bird can catch it. The predatory bird is the next host for the parasite in the next stage of its life cycle. Specifically, the tapeworm *Schistocephalus solidus* turns infected threespine stickleback white, and then makes them more buoyant so that they splash along at the surface of the water, becoming easy to see and easy to catch for a passing bird. Other parasitic disorders, include *Gyrodactylus salaris, Ichthyophthirius multifiliis*, cryptocaryon, velvet disease, *Brooklynella hostilis*, Hole in the head, *Glugea*, *Ceratomyxa shasta*, *Kudoa thyrsites*, *Tetracapsuloides bryosalmonae*, *Cymothoa exigua*, leeches, nematode, flukes, *Platyhelminthes*, carp lice and salmon lice

Mass Die Offs

Some diseases result in mass die offs. One of the more bizarre and recently discovered diseases produces huge fish kills in shallow marine waters. It is caused by the ambush predator dinoflagellate *Pfiesteria piscicida*. When large numbers of fish, like shoaling forage fish, are in confined situations such as shallow bays, the excretions from the fish encourage this dinoflagellate, which is not normally toxic, to produce free-swimming zoospores. If the fish remain in the area, continuing to provide nourishment, then the zoospores start secreting a neurotoxin. This toxin results in the fish developing bleeding lesions, and their skin flakes off in the water. The dinoflagellates then eat the blood and flakes of tissue while the affected fish die. Fish kills by this dinoflagellate are common, and they may also have been responsible for kills in the past which were thought to have had other causes. Kills like these can be viewed as natural mechanisms for regulating the population of exceptionally abundant fish. The rate at which the kills occur increases as organically polluted land runoff increases.

Cleaner Fish

Figure: *Two cleaner wrasses,* Labroides phthirophagus, *servicing a goatfish,* Mulloidichthys flavolineatus

Some fish take advantage of cleaner fish for the removal of external parasites. The best known of these are the Bluestreak cleaner wrasses of the genus *Labroides* found on coral reefs in the Indian Ocean and Pacific Ocean. These small fish maintain so-called "cleaning stations" where other fish, known as hosts, will congregate and perform specific movements to attract the attention of the cleaner fish. Cleaning behaviours have been observed in a number of other fish groups, including an interesting case between two cichlids of the same genus, *Etroplus maculatus*, the cleaner fish, and the much larger *Etroplus suratensis*, the host.

More than 40 species of parasites may reside on the skin and internally of the ocean sunfish, motivating the fish to seek relief in a number of ways. In temperate regions, drifting kelp fields harbour cleaner wrasses and other fish which remove parasites from the skin of visiting sunfish. In the tropics, the *mola* will solicit cleaner help from reef fishes. By basking on its side at the surface, the sunfish also allows seabirds to feed on parasites from their skin. Sunfish have been reported to breach more than ten feet above the surface, possibly as another effort to dislodge parasites on the body.

Wild Salmon

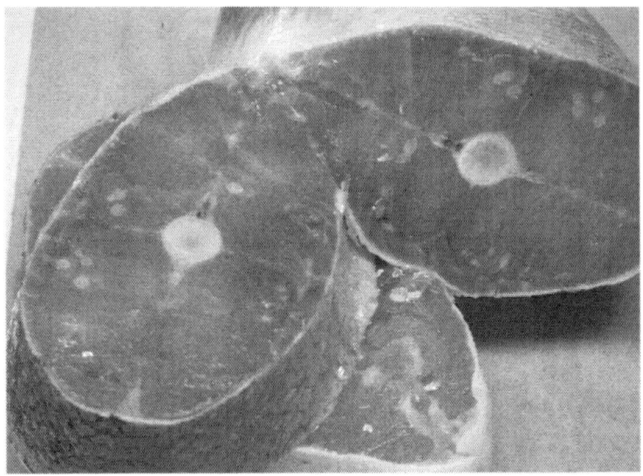

Figure: Henneguya salminicola, *a protozoan parasite commonly found in the flesh of salmonids on the West Coast of Canada. Coho salmon*

According to Canadian biologist Dorothy Kieser, protozoan parasite *Henneguya salminicola* is commonly found in the flesh of salmonids. It has been recorded in the field samples of salmon returning to the Queen Charlotte Islands. The fish responds by walling off the parasitic infection into a number of cysts that contain milky fluid. This fluid is an accumulation of a large number of parasites.

Henneguya and other parasites in the myxosporean group have a complex lifecycle where the salmon is one of two hosts. The fish releases the spores after spawning. In the *Henneguya* case, the spores enter a second host, most likely an invertebrate, in the spawning stream. When juvenile salmon out-migrate to the Pacific Ocean, the second host releases a stage infective to salmon. The parasite is then carried in the salmon until the next spawning cycle. The myxosporean parasite that causes whirling disease in trout, has a similar lifecycle. However, as opposed to whirling disease, the *Henneguya* infestation does not appear to cause disease in the host salmon — even heavily infected fish tend to return to spawn successfully.

According to Dr. Kieser, a lot of work on *Henneguya salminicola* was done by scientists at the Pacific Biological Station in Nanaimo in the mid-1980s, in particular, an overview report which states that "the fish that have the longest fresh water residence time as juveniles have the most noticeable infections. Hence in order of prevalence coho are most infected followed by sockeye, chinook, chum and pink." As well, the report says that, at the time the studies were conducted, stocks from the middle and upper reaches of large river systems in British Columbia such as Fraser, Skeena, Nass and from mainland coastal streams in the southern half of B.C. "are more likely to have a low prevalence of infection." The report also states "It should be stressed that *Henneguya*, economically deleterious though it is, is harmless from the view of public health. It is strictly a fish parasite that cannot live in or affect warm blooded animals, including man".

According to Klaus Schallie, Molluscan Shellfish Program Specialist with the Canadian Food Inspection Agency, "*Henneguya salminicola* is found in southern B.C. also and in all species of salmon. I have previously examined smoked chum salmon sides that were riddled with cysts and some sockeye runs in Barkley Sound (southern B.C., west coast of Vancouver Island) are noted for their high incidence of infestation."

Sea lice, particularly *Lepeophtheirus salmonis* and a variety of *Caligus* species, including *Caligus clemensi* and *Caligus rogercresseyi*, can cause deadly infestations of both farm-grown and wild salmon. Sea lice are ectoparasites which feed on mucous, blood, and skin, and migrate and latch onto the skin of wild salmon during free-swimming, planktonic *naupli* and *copepodid* larval stages, which can persist for several days. Large numbers of highly populated, open-net salmon farms can create exceptionally large concentrations of sea lice; when exposed in river estuaries containing large numbers of open-net farms, many young wild salmon are infected, and do not survive as a result.

Adult salmon may survive otherwise critical numbers of sea lice, but small, thin-skinned juvenile salmon migrating to sea are highly vulnerable. On the Pacific coast of Canada, the louse-induced mortality of pink salmon in some regions is commonly over 80%.

Farmed Salmon

Figure: *Atlantic salmon*

In 1972, Gyrodactylus, a monogenean parasite, spread from Norwegian hatcheries to wild salmon, and devastated some wild salmon populations.

In 1984, infectious salmon anemia (ISAv) was discovered in Norway in an Atlantic salmon hatchery. Eighty percent of the fish in the outbreak died. ISAv, a viral disease, is now a major threat to the viability of Atlantic salmon farming. It is now the first of the diseases classified on List One of the European Commission's fish health regime. Amongst other measures, this requires the total eradication of the entire fish stock should an outbreak of the disease be confirmed on any farm. ISAv seriously affects salmon farms in Chile, Norway, Scotland and Canada, causing major economic losses to infected farms. As the name implies, it causes severe anemia of infected fish. Unlike mammals, the red blood cells of fish have DNA, and can become infected with viruses. The fish develop pale gills, and may swim close to the water surface, gulping for air. However, the disease can also develop without the fish showing any external signs of illness, the fish maintain a normal appetite, and then they suddenly die. The disease can progress slowly throughout an infected farm and, in the worst cases, death rates may approach 100 percent. It is also a threat to the dwindling stocks of wild salmon. Management strategies include developing a vaccine and improving genetic resistance to the disease.

In the wild, diseases and parasites are normally at low levels, and kept in check by natural predation on weakened individuals. In crowded net pens they can become epidemics. Diseases and parasites also transfer from farmed to wild salmon populations. A recent study in

British Columbia links the spread of parasitic sea lice from river salmon farms to wild pink salmon in the same river." The European Commission (2002) concluded "The reduction of wild salmonid abundance is also linked to other factors but there is more and more scientific evidence establishing a direct link between the number of lice-infested wild fish and the presence of cages in the same estuary." It is reported that wild salmon on the west coast of Canada are being driven to extinction by sea lice from nearby salmon farms. Antibiotics and pesticides are often used to control the diseases and parasites.

Aquarium Fish

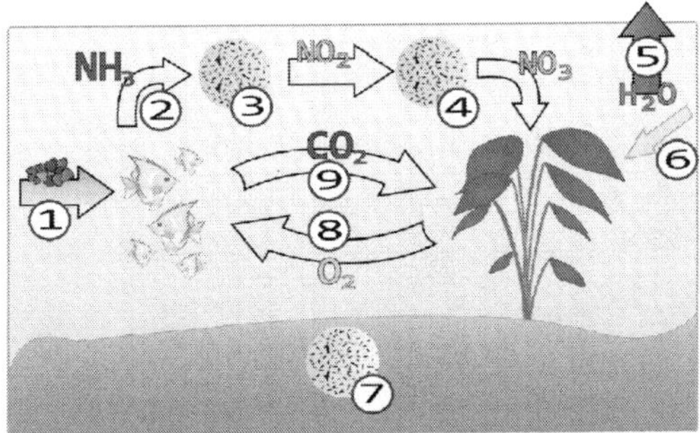

Figure: *Nitrogen cycle in a common aquarium.*

Ornamental fish kept in aquariums are susceptible to numerous diseases. In most aquarium tanks, the fish are at high concentrations and the volume of water is limited. This means that communicable diseases can spread rapidly to most or all fish in a tank. An improper nitrogen cycle, inappropriate aquarium plants and potentially harmful freshwater invertebrates can directly harm or add to the stresses on ornamental fish in a tank. Despite this, many diseases in captive fish can be avoided or prevented through proper water conditions and a well-adjusted ecosystem within the tank. Ammonia poisoning is a common disease in new aquariums, especially when immediately stocked to full capacity. Due to their generally small size and the low cost of replacing diseased or dead aquarium fish, the cost of testing and treating diseases is often seen as more trouble than the value of the fish.

Spreading Disease and Parasites

The capture, transportation and culture of bait fish can spread damaging organisms between ecosystems, endangering them. In 2007, several American states, including Michigan, enacted regulations designed to slow the spread of fish diseases, including viral hemorrhagic septicemia, by bait fish. Because of the risk of transmitting *Myxobolus cerebralis* (whirling disease), trout and salmon should not be used as bait. Anglers may increase the possibility of contamination by emptying bait buckets into fishing venues and collecting or using bait improperly. The transportation of fish from one location to another can break the law and cause the introduction of fish and parasites alien to the ecosystem.

Eating Raw Fish

Ornamental fish kept in aquariums are susceptible to numerous diseases. In most aquarium tanks, the fish are at high concentrations and the volume of water is limited. This means that communicable diseases can spread rapidly to most or all fish in a tank. An improper nitrogen cycle, inappropriate aquarium plants and potentially harmful freshwater invertebrates can directly harm or add to the stresses on ornamental fish in a tank. Despite this, many diseases in captive fish can be avoided or prevented through proper water conditions and a well-adjusted ecosystem within the tank. Ammonia poisoning is a common disease in new aquariums, especially when immediately stocked to full capacity. Due to their generally small size and the low cost of replacing diseased or dead aquarium fish, the cost of testing and treating diseases is often seen as more trouble than the value of the fish.

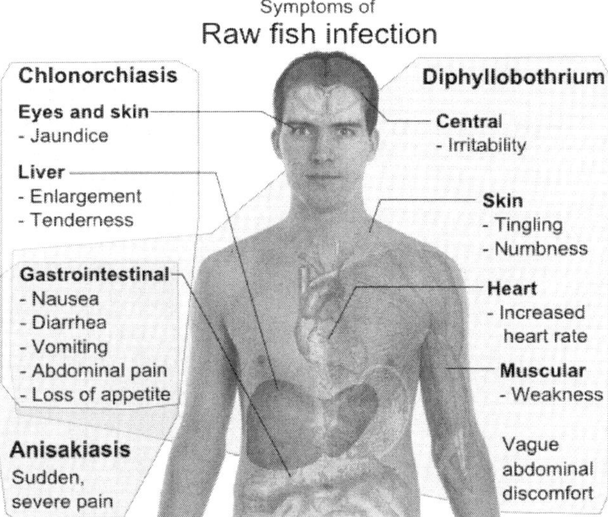

Figure: *Differential symptoms of parasite infection by raw fish: Clonorchis sinensis (a trematode/fluke), Anisakis (a nematode/ roundworm) and Diphyllobothrium a (cestode/tapeworm), all have gastrointestinal, but otherwise distinct, symptoms.*

Though not a health concern in thoroughly cooked fish, parasites are a concern when human consumers eat raw or lightly preserved fish such as sashimi, sushi, ceviche, and gravlax. The popularity of such raw fish dishes makes it important for consumers to be aware of this risk. Raw fish should be frozen to an internal temperature of "20°C ("4°F) for at least 7 days to kill parasites. It is important to be aware that home freezers may not be cold enough to kill parasites.

Traditionally, fish that live all or part of their lives in fresh water were considered unsuitable for sashimi due to the possibility of parasites. Parasitic infections from freshwater fish are a serious problem in some parts of the world, particularly Southeast Asia. Fish that spend part of their life cycle in brackish or freshwater, like salmon are a particular problem. A study in Seattle, Washington showed that 100% of wild salmon had roundworm larvae capable of infecting people. In the same study farm raised salmon did not have any roundworm larvae.

Parasite Infection by Raw Fish is Rare in the Developed World and Involves mainly three kinds of parasites: Clonorchis sinensis (a trematode/fluke), Anisakis (a nematode/roundworm) and Diphyllobothrium (a cestode/tapeworm). Infection risk of anisakis is particularly higher in fishes which may live in a river such as salmon (*shake*) in Salmonidae, mackerel (*saba*). Such parasite infections can generally be avoided by boiling, burning, preserving in salt or vinegar, or freezing overnight. Even Japanese people never eat raw salmon and ikura, and even if they seem raw, these foods are not raw but are frozen overnight to prevent infections from parasites, particularly anisakis. infections from parasites, particularly anisakis.

Aeromonas Salmonicida

Aeromonas salmonicida is a species of Gram-negative bacteria, from the genus Aeromonas, which causes the disease furunculosis in marine and freshwater fish.

Cell Structure and Metabolism

Aeromonas salmonicida is a gram-negative bacterium. Most strands of the bacterium are non-motile. It is bacillus in shape. The short rods have rounded ends which allow it to be easily confused as a coccus. *Aeromonas salmonicida* is a facultative anaerobe which means that it is capable of making ATP by aerobic respiration if oxygen is present but is also capable of switching to fermentation when oxygen is not present. It does not ferment sucrose or lactose, using glucose in this pathway instead; glucose fermentation creates gas. The bacterium grows optimally at temperatures between 22 and 25°C. The maximum temperature that it can grow at is 34.5°C. After about a 24 hour growth period the bacterial colonies reach about the size of a pin point. The colonies also have a brown pigmented color that appears after it has been growing for 48–72 hours.

Pathology

The bacterium is pathogenic for fish, and causes a disease known as furunculosis. The symptoms the fish show are external and internal hemorrhaging, swelling of the vents and kidneys, boils, ulcers, liquefaction, and gastroenteritis. Furunculosis is commonly known as tail rot in fish and is common in gold and koi fish. Infected fish with open sores are able to spread the disease to other fish. It is also responsible for bald sea urchin disease.

Detection

A. salmonicida tests negative for indole formation, coagulase, hydrolysis of starch, casein, triglycerides, and phospholipids, hydrogen sulfide production, citrate utilization, phenylalanine and the Voges–Proskauer (butanediol fermentation) test. It tests positive for oxidase, lysine decarboxylase, methyl red, gelatin hydrolysis, and catalase.

Columnaris

Figure: *Columnaris in a Chinook salmon*

Columnaris is a symptom of disease in fish which results from an infection caused by the gram negative, aerobic, rod shaped bacterium *Flavobacterium columnare*. It was previously known by the names of bacillus columnaris, chondrococcus columnaris, cytophaga columnaris and flexibacter columnaris. The bacteria is ubiquitous in fresh water, and cultured fish reared in ponds or raceways are the primary concern – with disease most prevalent in air temperatures above 12–14p C. The disease is highly contagious and the outcome is often fatal. It is not zoonotic.

Causes

The bacterium usually enters fish through gills, mouth, or small wounds, and is prevalent where high bio-load exists, or where conditions may be stressful due to overcrowding or low dissolved oxygen levels in the water column. The bacteria can persist in water for up to 32 days when the hardness is 50 ppm or more.

Symptoms

An infection will usually first manifest in fish by causing frayed and ragged fins. This is followed by the appearance of ulcerations on the skin, and subsequent epidermal loss, identifiable as white or cloudy fungus-like patches – particularly on the gill filaments. Mucus often also accumulates on the gills, head and dorsal regions. Gills will change colour, either becoming light or dark brown, and may also manifest necrosis. Fish will breathe rapidly and laboriously as a sign of gill damage. Inappetance and lethargy are common, as are mortalities – especially in young fish.

Diagnosis

Bacteria can be isolated from gills, skin and the kidneys. For definitive diagnosis the pathogen should then be cultured on reduced nutrient agar. Inhibiting contaminant growth on the agar by adding antibiotics and keeping the temperature at 37°C should improve culture results. Colonies are small, 3–4 mm in diameter and grow within 24 hours. They are characteristically rhizoid in structure and pale yellow in colour.

Prognosis

Ulcerations develop within 24 to 48 hours. Fatality occurs between 48 to 72 hours if no treatment is pursued; however, at higher temperatures death may occur within hours. Other symptoms may accompany the disease including lethargy, color loss, redness around the infection site, loss of appetite and twitching or rubbing the body against objects.

Treatment

As Flavobacterium columnare is a gram-negative bacteria, fish can be treated with a combination of the antibiotics Furan-2 and Kanamycin administered together. Medicated food containing oxytetracycline is also an effective treatment for internal infections, but resistance is emerging. Potassium permanganate, copper sulphate and hydrogen peroxide can also be applied externally to adult fish and fry but can be toxic at high concentrations. Vaccines can also be given in the face of an outbreak or to prevent disease occurrence.

Enteric Redmouth Disease

Enteric redmouth disease, or simply redmouth disease is a bacterial infection of freshwater and marine fish caused by the pathogen *Yersinia ruckeri*. It is primarily found in rainbow trout (*Oncorhynchus mykiss*) and other cultured salmonids. The disease is characterized by subcutaneous hemorrhaging of the mouth, fins, and eyes. It is most commonly seen in fish farms with poor water quality. Redmouth disease was first discovered in Idaho rainbow trout in the 1950s.. The disease does not affect humans.

Distribution of Disease

Some fish species serve as vectors for the disease and have subsequently spread the pathogen to other parts of the world. An example is the fathead minnow (*Pimephales promelas*) which is responsible for the spread of redmouth disease to trout in Europe. Other vectors include the goldfish (*Carassius auratus*), Atlantic and Pacific salmon (*Salmo salar*), the emerald shiner (*Notropis atherinoides*), and farmed whitefish (*Coregonus spp.*). Infections have also occurred in farmed turbot (*Scophthalmus maximus*), seabass (*Dicentrarchus labrax*), and seabream (*Sparus auratus*). It can now be found in North and South America, Africa, Asia and Australia as well as Europe.

Clinical Signs and Diagnosis

Infection can cause subcutaneous haemorrhage, that presents as reddening of the throat, mouth, gill tips and fins and eventual erosion of the jaw and palate. Hemorrhaging also occurs on internal organs, and in the later stages of the disease the abdomen becomes filled with a yellow fluid - giving the fish a "pot-bellied" appearance. The fish often demonstrate abnormal behavior and inappetance. Mortality rates can be high.

A presumptive diagnosis can be made based in the history and clinical signs, but definitive diagnosis requires bacterial culture and serological testing such as ELISA and latex agglutination.

Treatment and Control

There are several antibiotics available for the treatment of redmouth disease in fish. Vaccines can also be used in the treatment and prevention of disease. Management factors such as maintaining water quality and a low stocking density are essential for disease prevention.

Fin Rot

Fin rot is a symptom of disease or the actual disease in fish. This is a disease which is most often observed in aquaria and aquaculture, but can also occur in natural populations. Fin rot can be the result of a bacterial infection (*Pseudomonas fluorescens*, which causes a ragged rotting of the fin), or as a fungal infection (which rots the fin more evenly and is more likely to produce a white 'edge'). Sometimes, both types of infection are seen together. Infection is commonly brought on by bad water conditions, injury, poor diet, or as a secondary infection in a fish which is already stressed by other disease.

Fin rot starts at the edge of the fins, and destroys more and more tissue until it reaches the fin base. If it does reach the fin base, the fish will never be able to regenerate the lost tissue. At this point, the disease may attack the fish's body directly.

Very Common In:

- Betta fish; due to poor water conditions in pet stores.

Symptoms:

- Fin edges turn black / brown
- Fins fray
- Base of fins inflamed
- Entire fin may rot away or fall off in large chunks

Ways to treat:

- Change the water
- Treat with antibiotics like Melafix
- Use *aquarium salt* to keep water healthy
- Find out the ph and correct it
- Use Stress Coat

Use medication specifically for fin rot:

- Maracyn TC (for positive bacterial infections)
- Maracyn 2 (for negative bacterial infections)
- Melafix (as a preventative)

Prevention:

- Make sure the water is good quality
- Feed fresh food in small portions
- Maintain constant water temperature

Fish Dropsy

Figure: *A Goldfish with Fish Dropsy*

Dropsy is a common disease among fresh-water aquarium fish. It is characterized by a swollen or hollow abdomen. The name is from an old name for Edema in humans.

Symptoms

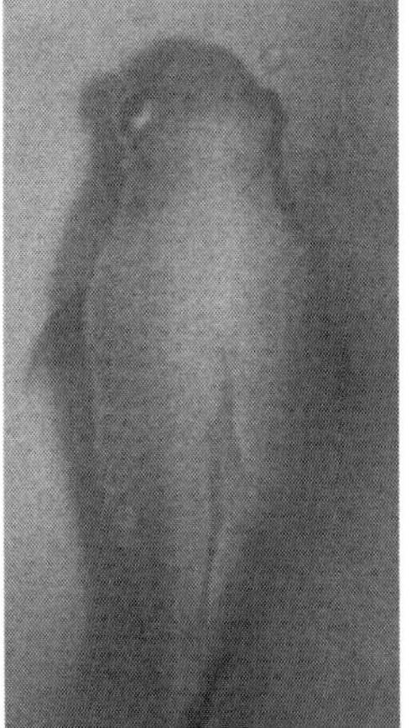

Figure: *A Gold Fish with Fish Dropsy in containment. Notice the pinecone appearance*

In dropsy, a concentration of fluid in the body tissues and cavities causes the fish's abdomen to become swollen and appear bloated (Ascites). Swollen areas may exhibit a 'pine-cone' appearance caused by the fish's scales sticking out. Fish may also stop feeding, appear off-color, become listless and/or lethargic, have sunken eyes, and hang at the top or stay at the bottom of the aquarium. The condition affects the fish's internal organs, ceasing proper function.

Causes

The cause of fish dropsy can be difficult to diagnose. The main cause is bacterial infection. The causative agent may be introduced through poor water quality. Kidney failure or excess fluid (ascites) due to liver or heart failure are other possible causes.

Treatment

Because dropsy is a symptom of an illness, it may or may not be contagious. However, it is standard practice to quarantine sick fish to prevent stress among the other fish in the tank community. This extra stress may make the others vulnerable to dropsy or other forms of disease.

Prognosis

Most cases of dropsy are fatal if not treated correctly. By the time the fish has swollen up enough that the scales begin to rise, the internal damage may be too extensive to repair.

Prevention

Water quality is an important factor in prevention of fish disease. Water changes will dilute existing disease agents, and reduce stress on the tank occupants.

Flavobacterium

Flavobacterium is a genus of Gram-negative, non-motile and motile, rod-shaped bacteria that consists of ten recognized species, as well as three newly proposed species (*F. gondwanense*, *F. salegens*, and *F. scophthalmum*). Flavobacteria are found in soil and fresh water in a variety of environments. Several species are known to cause disease in freshwater fish.

Flavobacterium psychrophilum causes the Bacterial Cold Water Disease (BCWD) on salmonids and the Rainbow Trout Fry Disease (RTFS) on rainbow trouts. *Flavobacterium columnare* causes the cotton-wool disease on freshwater fishes. *Flavobacterium branchiophilum* causes the Bacterial Gill Disease (BGD) on trouts.

Infectious Hematopoietic Necrosis Virus

Infectious hematopoietic necrosis virus (IHNV), is a negative-sense single-stranded, bullet-shaped RNA virus that is a member of the *Rhabdoviridae* family, and from the genus *Novirhabdovirus*. It causes the disease known as infectious hematopoietic necrosis in salmonid fish like trout and salmon.

The disease may be referred to by a number of other names such as Chinook Salmon Disease, Coleman Disease, Columbia River Sockeye Disease, Cultus Lake Virus Disease, Oregon Sockeye Disease, Sacramento River Chinook Disease and Sockeye Salmon Viral Disease. IHNV is commonly found in the Pacific Coast of Canada and the USA, and has also been found in Europe and Japan.

The first reported epidemics of IHNV occurred in the United States at the Washington and the Oregon fish hatcheries during the 1950s. IHNV is transmitted following shedding of the virus in the feces, urine, sexual fluids, and external mucus and by direct contact or close contact with surrounding water. The virus gains entry into fish at the base of the fins.

The disease is listed as a non-exotic disease of the EU and is therefore watched closely by the European Community Reference Laboratory for Fish Diseases. In order to keep track of the distribution of different IHNV genotypes a database called Fishpathogens.eu has been created to store data on different fish pathogens (including IHNV) and their sequences.

Classification

IHNV is the causal agent of Infectious Hematopoietic Necrosis (IHN) disease of fish and is a negative-sense single-stranded RNA virus. IHNV is a member of the genus *Novirhabdovirus*, belongs to the family of *Rhabdoviridae*. The North American IHNV isolates are grouped based on the partial Glycoprotein (G) gene sequences. There are 3 major IHNV genogroups exist in North America and are designated as U, M and L for the upper, middle and lower portions of IHNV geographical range in North America. The Japanese and Korean isolates constitute new JRt (Japanese Rainbow trout) genogroup.

Virion

Virions consist of an envelope and a nucleocapsid. Virions are bullet-shaped and measure 45-100 nm in diameter; 100-430 nm in length. Surface projections are densely dispersed, distinctive spikes that cover the whole surface except for the quasi-planar end.

Genome Organization

The fish rhabdovirus, IHNV, has a bullet-shaped virion containing a non-segmented, negative-sense, single-stranded RNA genome of approximately 11,000 nucleotides that encodes six proteins in the following order: a nucleoprotein (N), a phosphoprotein (P), a matrix protein (M), a glycoprotein (G), a nonvirion protein (NV), and a polymerase (L). To date,complete genome sequence has been available for only 3 IHNV isolates.

IHNV Replication Cycle

The rhabdoviral cycle of infection occurs by series of events in the following order: adsorption, penetration and uncoating, transcription, translation, replication, assembly and budding.

Transmission

Reservoirs of IHNV are clinically infected fish and covert carriers among cultured, feral or wild fish. Virus is shed via urine, sexual fluids and from external mucus, whereas kidney, spleen and other internal organs are the sites in which virus is most abundant during the course of overt infection. Insect, annelids and crustaceans can act as viral vectors. The potential for epizootics is highest at 10°C and the disease does not occur naturally above 15°C.

Clinical Signs

Clinical signs of infection with IHNV include abdominal distension, bulging of the eyes, skin darkening, abnormal behavior, anemia and fading of the gills. Infected fish commonly hemorrhage in several areas; the mouth and behind the head, the pectoral fins, muscles near the anus, and (in fry) the yolk sac. Diseased fish weaken eventually floating “belly-up” on the surface of the water.

Necrosis is common in the kidney and spleen, and sometimes in the liver. Mortality is very high in young fish. Some fish become covert carriers of the vrius if they survive infection.

Diagnosis

Clinical signs and history of previous outbreaks may be suggestive of IHN. Staphylococcal agglutination, virus neutralisation (VN), indirect fluorescent antibody testing, ELISA, PCR and DNA probe technology are all techniques that can be used to confirm diagnosis. The gold standard is virus neutralisation. Alternatively, the identification of degeneration and necrosis of granular cells in the lamina propria, stratum compactum and stratum granulosum of the gastrointestinal tract on histopathology can be used to diagnose infection.

Treatment and Control

No treatment had yet proven to be effective. To prevent the disease, strict isolation, hygiene and testing procedures should be in place.

Heterosigma Akashiwo

Heterosigma akashiwo is a microscopic alga of the class Raphidophyceae. *Heterosigma akashiwo* is a swimming marine alga that episodically forms toxic surface aggregations known as harmful algal bloom (HAB).

Etymology

The species name "*akashiwo*" is from the Japanese for "red tide."

Context and Content

Synonymns include *Olisthodiscus luteus* (Hulburt 1965), and *Entomosigma akashiwo* (Hada 1967). *H. akashiwo* and *H. inlandica* have been recognized as two species of *Heterosigma*. However, Hara and Chihara (1987) described both specimens as one species, validly describing them as *H. akashiwo*.

Description

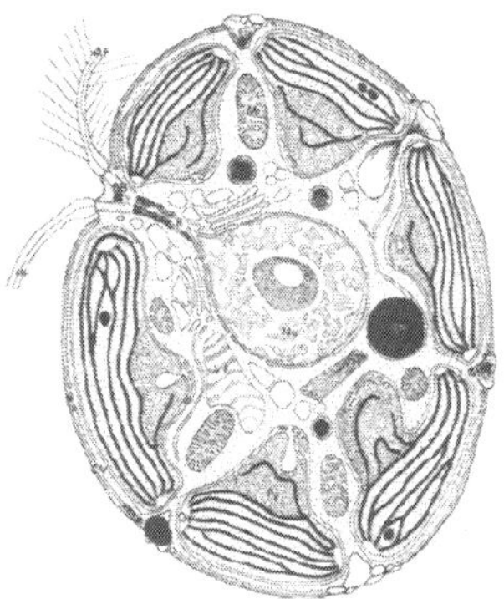

Figure: Heterosigma akashiwo *anatomy from Hara and Chihara 1987*

H. akashiwo cells are relatively small, ranging in size from 18 to 34 um in diameter. They appear golden brown, and appear in clusters. Morphology is highly variable, but does not appear to vary significantly between locations. One culture may contain individual cells that are flat or round. . Molecular techniques for identification (including real-time PCR) are preferred over traditional microscope fixing, which may lyse the cells.

Distribution

H. akashiwo has been identified off the coasts of The United States, Canada, Chile, the Netherlands, Scotland, Ireland, Sweden,

Norway, Japan, Australia, and New Zealand. Most of the literature suggests that *H. akashiwo* is associated with shallow water within 10 m of the surface, but this is not a universal rule.

Physiology

H. akashiwo is a mixotrophic alga, supplementing nutrient uptake and photosynthesis with ingestion of bacteria. Each cell may contain 18-27 chloroplasts. These cells have been observed to glide and twirl under microscopic examination, but non-motile cells have been associated with toxic blooms. Blooms are clearly visible by air, appearing as a red area in otherwise blue water. Optimal growth occurs at 25°C and 100 ìE ms_{-1}, conditions which are associated with very low toxicity. Maximum toxicity occurs (and relatively slow growth) occurs at 20°C and 200 ìE ms_{-1}. *H. akashiwo* reproduces asexually by binary fission.

Heterosigma akashiwo produces cysts as a resting stages. The germination of these cysts leads to large-scale blooms, which can be laterally transferred by tides and currents. These blooms do not seem to be caused by fish farming. Bottom water temperature must reach at least 15°C for germination to occur. Blooms are associated with summer months, and some areas may see two blooms within one year. Blooms are known to be lethal once concentrations of cells reach 3x10 to 7 x 10^5 cells/L. Viruses may act as a natural control on bloom populations, as *H. akashiwo* viruses (HaV) have been shown to only leave resistant alga alive. Similarly, certain bacteria may also reduce *H. akashiwo* populations.

The exact mode of bloom toxicity is currently unknown, but gill damage leading to hypoxia is the proposed cause for fish death. *H. akashiwo* may produce brevetoxins, but others suggest that concentrations of these toxins are too low to account for such a large effect on fish populations. Some have argued that the production of reactive oxygen species like hydrogen peroxide may be responsible for gill damage. However, research suggests that hydrogen peroxide concentrations are far too low to have significant effects on fish. Mucus production is another proposed, but poorly supported, mechanism for fish mortality. It is possible that the effective toxin is chemically unstable, and therefore difficult to detect. Sablefish appear to be unaffected by *H. akashiwo* blooms, while many other marine fish are decimated.

Genetics

Genetic sequences are highly conserved between Pacific and Atlantic populations. Relevant probe sequences for small subunit RNA can be found in Bowers et al. 2006.

Economic Impact

Heterosigma forms massive golden tides that impact the survival of organisms at every trophic level. This alga has been shown to kill finfish, compromise fish and sea urchin egg development and impact copepod as well as oyster survival. Further ecological impacts to plankton, invertebrates, and wild fish are likely, but unknown. The 1997 *H. akashiwo* bloom in British Columbia, for example, coincided with a dramatic increase in mortality of captive salmon. *H. akashiwo* contributed to the loss of over 1,000 tons of Atlantic salmon in 2001. A bloom in Puget Sound in 2006 led to the loss of $2 million of farmed salmon. Moreover, the global distribution of *H. akashiwo* is increasing as is the frequency of *H. akashiwo* HAB formation.

Head and Lateral Line Erosion

Figure: *A fish that has died from HLLE*

Head and lateral line erosion, also known as HLLE, hole-in-the-head disease, or HITH, is a fish disease that affects both freshwater and marine fish in captivity. Sometimes marine head and lateral line erosion or MHLLE is used to refer to the marine version, though the causes and treatments for each are similar. Among freshwater fish, it affects cichlids and angelfish, with Oscars developing the disease more frequently than other related fish. Angelfish, tangs, and groupers, and to a lesser degree lionfish, damselfish, and clownfish are susceptible in marine aquaria.

Symptoms

Beginning as small pits around the fish's eyes, light colored lesions then develop along the fish's lateral line system, onto the body and sometimes onto the unpaired fins. Rarely fatal, it does cause disfigurement, making the fish less suitable for public aquarium display. At least 20 families of fish have been identified as having developed HLLE in captivity. Not all species of fish show the same symptoms, and do not always develop lesions to the same degree (Hemdal 2006).

Causes

There is much debate as to the cause of HLLE. Very little scientific research has been done on the topic, and most information available is anecdotal. Removing carbon from the filter is a commonly suggested remedy for the disease. It is suggested that the carbon may either a.) add fine carbon particles to the water that irritate the skin, b.) leach potentially harmful chemicals into the water, or c.) remove minerals that are important to the health of the fish. Though there have been many reports of fish being affected by carbon, only one study has substantiated this; http://www.coralmagazine-us.com/content/hlle-and-activated-carbon-looking-link

Some people believe that parasites of the genus *Hexamita*, a flagellated protozoan, is to blame for this syndrome in freshwater fishes. HLLE and *Hexamita* infections are often seen in the same specimens. However, HLLE can be found in many fish who do not have a *Hexamita* infection, suggesting that a *Hexamita* infection may cause stress or interfere with the absorption of vitamins and minerals causing malnutrition, which may be the actual cause of the disease. *Hexamita* may also be a secondary infection common in fish already weakened by HLLE.

One non-replicated study showed that improvement in nutrition will help symptoms most effectively (Blasiola 1992). Key vitamins in preventing or curing HLLE seem to be Vitamins A, C, and D. Adding these vitamins to the diet of affected fish usually leads to improvement, however, deficiencies in any of these vitamins do not always lead to HLLE, so nutrition is also questionable as a cause.

It has also been suggested that HLLE is an autoimmune disorder that is triggered by stress. The disease does not appear to occur to fish in the wild, only those in captivity, supporting the idea that stress and unnatural living conditions are to blame. Anything that reduces stress appears to help in the prevention and recovery from this disease.

Infectious Hypodermal and Hematopoietic Necrosis

Infectious hypodermal and hematopoietic necrosis (IHHN) is a viral disease of penaeid shrimp that causes mass mortality (up to 90%)

among the Western blue shrimp (*Penaeus stylirostris*) and severe deformations in the Pacific white shrimp (*P. vannamei*). It occurs in Pacific farmed and wild shrimp, but not in wild shrimp on the Atlantic coast of the Americas. The shrimp-farming industry has developed several broodstocks of both *P. stylirostris* and *P. vannamei* that are resistant against IHHN infection. The disease is caused by a single-stranded DNA virus simply called "IHHN virus", the smallest of the known penaeid shrimp viruses (22 nm).

Virology

This virus has been classified as Penaeus stylirostris densovirus. The genome is 4.1 kilobases in length. There are three open reading frames in its genome: a left non structural protein (NS1) of 2001 base pairs (bp), a mid non structural protein (NS2) of 1092 bp and a right capsid protein of 990 bp. NSI appears to be a DNA helicase with ATPase activity. The mutation rate in the genome has been estimated to be $\sim 1.4 \times 10^{-4}$ substitutions/site/year. This is considerably higher than the rate in the double stranded DNA viruses and similar to that found in RNA viruses. It is also similar to the rate found in other single stranded DNA viruses.

Infectious Pancreatic Necrosis

Infectious pancreatic necrosis (IPN) is a severe viral disease of salmonid fish. It is caused by Infectious Pancreatic Necrosis Virus, which is a member of the birnaviridae family. This disease mainly affects young salmonids, such as trout or salmon, of less than 6 months, although adult fish may carry the virus without showing symptoms. Resistance to infection develops more rapidly in warmer water. It is highly contagious and found worldwide, however some regions have managed to eradicate or greatly reduced the incidence of disease. The disease is normally spread horizontally via infected water but spread also occurs vertically. It is not a zoonosis.

Clinical Signs and Diagnosis

A sharp rise in mortality is often seen (depending on the virulence of the disease). Other clinical signs include abdominal swelling, anorexia, abnormal swimming, darkening of the skin and trailing of the feces from the vent. On postmortem there is often internal damage (viral necrosis) to the pancreas and thick mucus in the intestine.

Suriviving Fish should Recover within 1-2 Weeks

Diagnostic methods for the detection of the disease include: characteristic histological pancreatic lesion, PCR, indirect fluorescent antibody testing, ELISA and virus culture. High virus titers can be isolated from carrier animals.

Treatment and Control

There is currently no treatment: Good husbandry measures such as high water quality, low stocking density and no mixing of batches help to reduce disease incidence. To eradicate the disease very strict protocol with regards to movement, water sources and stock replacement must be in place - and still it is difficult to achieve and comes at a high economic cost.

Koi Herpes Virus

Koi herpes virus (KHV) is a viral disease that is very contagious to the common carp *Cyrpinus carpio*. It is most common found in ornamental koi, which are often used in outdoor ponds or as feeder stock. The first case of KHV was reported in 1998, but not confirmed until later in 1999.

KHV is a DNA-based virus. After discovery, it was identified that KHV is indeed a strain of herpesvirus. Like other strains, KHV stays with the infected fish for the duration of their life, making the recovered and exposed fish potential carriers of the virus. Koi fish infected with KHV may die within the first 24-48 hours of exposure.

KHV is listed as a non-exotic disease of the EU and is therefore watched closely by the European Community Reference Laboratory for Fish Diseases.

Symptoms of KHV

Symptoms of KHV include:

- Gill mottling
- Red and white patches appearing on gills
- Bleeding gills
- Sunken eyes
- Pale patches
- Blisters

Novirhabdovirus

Novirhabdovirus is one of the genera of Rhabdoviridae, along with *Vesiculovirus*, known to infect aquatic hosts. They can be transmitted from fish to fish, by waterborne virus, as well as through contaminated eggs. Replication and thermal inactivation temperatures are generally lower than for other rhabdoviruses, given the cold-blooded nature of their hosts. Hosts include a large and growing range of marine and freshwater fish. A common characteristic among novirhabdoviruses is the NV gene, an approximately 500 nucleotide long gene located between the glycoprotein (G) and polymerase (L) genes. The expected protein encoded by the NV gene is not found in the virions, leading to its being named a "nonvirion" (NV) protein. This is the origin of the genus name *Novirhabdovirus*.

The four recognized species of the genus are *Hirame rhabdovirus*, currently isolated to Japan, *Infectious hematopoietic necrosis virus* (IHNV), enozootic to North America but now present in Europe and Asia as well, *Viral hemorrhagic septicemia virus* (VHSV), enzootic to Europe but now present in North America and Asia as well, and *Snakehead rhabdovirus* (SHRV). Two tentative species of the genus, identified the Seventh Report of the ICTV, are *Eel virus B12* (EEV-B12), and *Eel virus C26* (EEV-C26).

Pfiesteria Piscicida

Pfiesteria piscicida is a dinoflagellate species of the genus *Pfiesteria* that some researchers claim is responsible for many harmful algal blooms in the 1980s and 1990s on the coast of North Carolina and Maryland. The species epithet *piscicida* means "fish-killer."

Life Cycles

Early research suggested a very complex life cycle of *Pfiesteria piscicida* with up to 24 different stages, spanning from cyst to several amoeboid forms with toxic zoospores. Transformations from one stage to another depend on environmental conditions such as the availability of food. However these results have become controversial as additional research has found only a simple haplontic life cycle with no toxic amoeboid stages and amoeba present on attacked fish may represent an unrelated species of protist.

Toxicity

Pfiesteria presumably kills fish via releasing a toxin into the water to paralyze its prey. This hypothesis has been questioned as no toxin could be isolated and no toxicity was observed in some experiments. However, toxicity appears to depend on the strains and assays used. Polymerase chain reaction-analyses suggested that the organism lacks the DNA genes for polyketide synthesis, the type of toxins associated with most toxic dinoflagellates. Researchers from the NOAA National Centers for Coastal Ocean Science, the National

Institute of Standards and Technology, the Medical University of South Carolina, and the College of Charleston (S.C.) have formally isolated and characterized the toxin in the estuarine dinoflagellete *Pfiesteria piscicida* as a metal complex and free radical toxin and also have identified how the organism transforms from a non-toxic to toxic state.

Human Health

Very little research on the human health effects of *Pfiesteria* toxins has been conducted. At a multi-state workshop at the Centers for Disease Control and Prevention (CDC) in Atlanta, U.S., at the end of September 1997, attendees agreed on clinical symptoms that characterize the adverse health consequences of exposure to *Pfiesteria* toxins. These clinical features include:

- memory loss
- confusion
- acute skin burning (on direct contact with water); or
- three or more of an additional set of conditions (headaches, skin rash, eye irritation, upper respiratory irritation, muscle cramps, and gastrointestinal complaints (i.e., nausea, vomiting, diarrhea, and/or abdominal cramps).

With these criteria and environmental qualifiers (e.g., 22% of a 50-fish sample, all of the same species, have lesions caused by a toxin), it is likely that *Pfiesteria*-related surveillance data can better track potential illnesses.

Pfiesteria toxins have been blamed for causing adverse health effects in people who have come in close contact with waters where this organism is abundant. Since June 1997, the Maryland Department of Health and Hygiene has been collecting data from Maryland physicians through a state-wide surveillance system on illnesses suspected of being caused by *Pfiesteria* toxin. As of late October 1997, illness was reported by 146 persons who had been exposed to diseased fish or to waters that were the site of suspected *Pfiesteria* activity. Many of these persons are watermen and commercial fishermen.

The strongest evidence of adverse human health effects so far comes from case studies of two research scientists who were both overcome in their North Carolina laboratory in 1993. They still complain of adverse effects on their cognitive abilities, particularly after exercising. Duke University Medical Center researchers conducted experiments on rats, which showed that the toxin appeared to slow learning but did not affect memory.

Photobacterium Damselae Subsp. Piscicida

Photobacterium damselae subsp. *piscicida* (previously known as *Pasteurella piscicida*) is a gram-negative rod-shaped bacterium that causes disease in fish.

Disease

Pasteurellosis is also described as photobacteriosis (due to the change in the taxonomic position), is caused by the halophilic bacterium *Photobacterium damselae* subsp. *piscicida* (formerly *Pasteurella piscicida*). It was first isolated in mortalities occurring in natural populations of white perch (*Morone americanus*) and striped bass (*M. saxatilis*) in 1963 in Chesapeake Bay, USA (*Snieszko* et al., *1964*). Since 1969, this disease has been one of the most important in Japan, affecting mainly yellowtail (*Seriola quinqueradiata*) (Kusuda & Yamaoka, 1972). From 1990 it has caused economic losses in different European countries including France (Baudin-Laurencin *et al.*, 1991), Italy (Ceschia *et al.*, 1991), Spain (Toranzo *et al.*, 1991), Greece (Bakopoulos *et al.*, 1995), Turkey (Canand *et al.*, 1996), Portugal (Baptista *et al.*, 1996) and Malta (Bakopoulos *et al.*, 1997). Gilthead sea bream (*Sparus aurata*), seabass (*Dicentrarchus labrax*) and sole (*Solea* spp.) are the most affected species in Europe Mediterranean countries, as well as hybrid striped bass (*M. saxatilis* x *M. chrysops*) in the USA. However, the natural hosts of the pathogen are a wide variety of marine fish (Romalde & Magariños, 1997).

This pathology is temperature dependent and occurs usually when water temperatures rise above 18-20°C. Below this temperature, fish can harbour the pathogen as subclinical infection and become carriers for long time periods (Romalde, 2002).

Symptoms

Pastereullosis is also known as pseudotubercullosis because it is characterized by the presence, in the chronic form of the disease, of creamy-white granulomatous nodules or whitish tubercules in several internal organs, composed of masses of bacterial cells, epithelial cells, and fibroblasts. The nodules are most prominent in internal viscera, particularly kidney and spleen, and the infection is accompanied by widespread internal necrosis (Evelyn, 1996; Romalde, 2002; Barnes *et al.*, 2005). Anorexia with darkening of the skin as well as focused necrosis of the gills are the only external clinical signs often observed. These lesions are generally missing in the acute form. The disease is difficult to eradicate with antibiotic treatments, and there is evidence that carriers under stressful conditions could suffer from reinfection (Le Breton, 1999).

Identification

Morphologically, the bacteria is a rod shaped cell, with no motility. Gram negative, with bipolar staining. The presumptive identification of the pathogen is based on standard biochemical tests. In addition, although *Ph. damselae* subsp. *piscicida* is not included in the API-20E code index, this miniaturised system can also be useful for its identification, since all strains display the same profile (2005004). Slide agglutination test using specific antiserum is needed for a confirmative identification of the microorganism (Romalde, 2002).

Virulence

The virulence of the pathogen implies the production of polysaccharide capsular layer, and extracellular products, and is also depending on iron availability (Lopez-Doriga *et al.*, 2000). The bacteria spreads via infected phagocytes, mainly macrophages. This spread can be rapid, and lethal effects may occur within a few days of challenge, affecting tissues containing large numbers of the pathogens (Evelyn, 1996).

7

Infectious Salmon Anemia Virus

Infectious salmon anemia or anaemia (ISA) is a viral disease of Atlantic Salmon (*Salmo salar*) that affects fish farms in Canada, Norway, Scotland and Chile, causing severe losses to infected farms. The disease is listed as a non-exotic disease of the EU and is therefore watched closely by the European Community Reference Laboratory for Fish Diseases.

The aetiological agent of ISA is the infectious salmon anemia virus (ISAV). ISAV, a RNA virus, is the only species in the genus "Isavirus" which is in the family Orthomyxoviridae.

Pathology

As the name implies, it causes severe anemia of infected fish. Unlike mammals, the red blood cells of fish have DNA, and can become infected with viruses. The fish develop pale gills, and may swim close to the water surface, gulping for air. However, the disease can also develop without the fish showing any external signs of illness, the fish maintain a normal appetite, and then they suddenly die. The disease can progress slowly throughout an infected farm and, in the worst cases, death rates may approach 100%. Post-mortem examination of the fish has shown a wide range of causes of death. The liver and spleen may be swollen, congested or partially already dead. The circulatory system may stop working, and the blood may be contaminated with dead blood cells. Red blood cells still present often burst easily and the numbers of immature and damaged blood cells are increased.

Infectious salmon anemia appears to be most like influenza viruses. Its mode of transfer and the natural reservoirs of infectious salmon anemia virus are not fully understood. Apart from Atlantic salmon, both sea-run Brown trout (*Salmo trutta*) and Rainbow trout (*Onchorhyncus mykiss*) can be infected, but do not become sick, so it is thought possible that these species may act as important carriers and reservoirs of the virus.

Epidemiology

Prevalence: In the autumn of 1984, a new disease was observed in Atlantic salmon being farmed along the southwest coast of Norway. The disease, which was named Infectious salmon anemia, spread slowly. By June 1988 it had become sufficiently widespread and serious to require the Norwegian Ministry of Agriculture, Fisheries and Food to declare it a notifiable disease.

In the summer of 1996, a new disease appeared in Atlantic salmon being farmed in New Brunswick, Canada. The death rate of the fish on affected farms was very high and, following extensive scientific examination of the victims, the disease was named "hemorrhagic kidney syndrome." Although the source and distribution of this disease was not known, the results of studies by Norwegian and Canadian scientists showed conclusively that the same virus was responsible for both infectious salmon anemia and hemorrhagic kidney syndrome.

In May 1998, a salmon farm at Loch Nevis on the west coast of Scotland reported its suspicions of an outbreak of infectious salmon anemia. The suspicions were confirmed, and by the end of the year, the disease had spread to an additional fifteen farms not only on the Scottish mainland but also on Skye and Shetland.

More recently (2008) there has been another outbreak of ISA in Shetland. ISA was detected in fish from just one site, which has since been harvested and will remain fallow. There is no evidence the disease has spread beyond this site, but two nearby SSF cages are under suspicion of carrying the disease and are also now clear of fish.

Most recently (2011) two wild pacific salmon were confirmed to have the disease off the central coast of British Columbia. According to the federal Centers for Disease Control and Prevention, infectious salmon anemia virus morphed from a benign form in nature into a "novel virulent strain" when salmon stocks entered Norway's densely packed salmon farms. Rather than getting picked off by a predator, a sick fish would undergo a slow death in a crowded pen, shedding virus particles.

In Chile, ISA was first isolated from a salmon farm in the 1990s and described for the first time in 2001 although the initial presence never resulted in widespread problems. However, since June 2007, the national industry has been dealing with a serious ISA outbreak which has not yet been completely brought under control and has been responsible for an important decline in the industry, closure of many farms and high unemployment. The virus was detected in an Atlantic salmon farm in Chiloé Archipelago in Los Lagos Region and spread to the fjords and channels of Aysén Region to the south almost immediately

Transmission

Transmission of the virus has been demonstrated to occur by contact with infected fish or their secretions. Contact with equipment or people who have handled infected fish also transmits the virus. The virus can survive in seawater and, not surprisingly, a major risk factor for any uninfected farm is its proximity to an already infected farm.

More recently the sea louse, a small crustacean parasite that attacks the protective mucous, scales and skin of the salmon has been shown to carry the virus passively on its surface and in its digestive tract, although transmission of the disease by sea lice has not been demonstrated. It is not known whether the Infectious salmon anemia virus can reproduce itself in the sea louse, although this is a possibility as viruses often depend on secondary vectors for transmission like the Arboviruses such as dengue fever, West Nile virus, or African swine fever virus.

Diagnosis

Clinical signs and pathology may suggest infection. Viral identification is possible using immunofluorescence and PCR.

Treatment and Control

There is no treatment once fish are infected.ISA is a major threat to the viability of salmon farming and is now the first of the diseases classified on List One of the European Commission's fish health regime. Amongst other measures, this requires the total eradication of the entire fish stock should an outbreak of the disease be confirmed on any farm. The economic and social consequences of both the disease and the measures used to control it are thus very far reaching.

Infectious salmon anemia is currently regarded as a serious threat not only to farmed salmon, but also to dwindling stocks of wild salmon.

Anecdotal evidence suggests that fish which survive the first infection become immune to the virus. Work is now underway to develop a vaccine against ISA. A recent report suggests that the North American virus may be slightly different from the Norwegian virus. This makes it unlikely that the sudden appearance of the disease, at least in Canada, was due to the importation of infected Norwegian fish. The possibility then is that a single vaccine might only be effective in a limited area and maybe only for a limited time.

Streptococcus Iniae

Streptococcus iniae is a species of Gram-positive, sphere-shaped bacterium belonging to the genus *Streptococcus*. Since its isolation from an Amazon freshwater dolphin in the 1970s, *S. iniae* has emerged as a leading fish pathogen in aquaculture operations worldwide, resulting in over US$100M in annual losses. Since its discovery, *S. iniae* infections have been reported in at least 27 species of cultured or wild fish from around the world. Freshwater and saltwater fish including tilapia, red drum, hybrid striped bass, and rainbow trout are among those susceptible to infection by *S. iniae*. Infections in fish manifest as meningoencephalitis, skin lesions, and septicemia.

S. iniae has occasionally produced infection in humans, especially fish handlers of Asian descent. Human infections include sepsis, toxic shock syndrome, and inflammation of the skin, intervertebral discs, or inner layer of the heart. Identifying *S. iniae* in the laboratory can be difficult, since the conventional methods used to identify streptococci yield insufficient results. It cannot be grouped by the Lancefield antigen method typically used to categorize *Streptococci* species. The two known serotypes can be distinguished biochemically by differences in enzyme activity. Several antibiotics have been used to treat *S. iniae* infections.

History

Streptococcus iniae was first isolated in 1972, from subcutaneous abscesses in a captive specimen of Amazon River Dolphin (*Inia geoffrensis*) suffering from an infection known as "golf ball disease". The bacterium was found to be sensitive to beta-lactam antibiotics, and the dolphin was treated successfully with penicillin and tylosin. The causative organism was recognized to be a new species of *Streptococcus*, and was given the name *Streptococcus iniae* in 1976. Around this time, other streptococcal outbreaks occurred in Asia, and the US; some of the strains associated with the Japanese outbreaks were later suggested to be *S. iniae*.

In the 1980s, a purported new species of *Streptococcus*, named *S. shiloi*, was identified as one of the causes of an epidemic of meningoencephalitis (an inflammation of the brain and its surrounding membranes) affecting farmed rainbow trout and tilapia in Israel since 1986. Since *S. shiloi* was alpha-hemolytic, had a G+C% content of 37% and did not ferment sugar galactose, it was not classified as *S. iniae*, which is beta-hemolytic, has a G+C% content of 32%, and ferments galactose. In 1995, *S. shiloi* was found in fact to be beta-hemolytic, and after DNA-DNA hybridization techniques with the ATCC type *S. iniae* and recalculation of the G+C% content, was reclassified by the same group as a junior synonym of *S. iniae*.

Phylogenetic analyses based on 16S ribosomal DNA suggest that *S. iniae* is closely related to other streptococcal pathogens of humans and animals. Specifically, it is clustered in the pyogenic group, along with other pathogenic streptococci such as *S. pyogenes*, *S. agalactiae*, *S. uberis*, *S. canis*, *S. porcinus*, *S. phocae*, and *S. intestinalis*. Of these related species, it is most closely related to *S. porcinus*. Genomic restriction fragment analysis of diverse host and geographical panels of *S. iniae* isolates has shown common profiles between virulent fish and human strains, though multiple pulsed field gel electrophoresis patterns have been identified among human isolates.

Identification

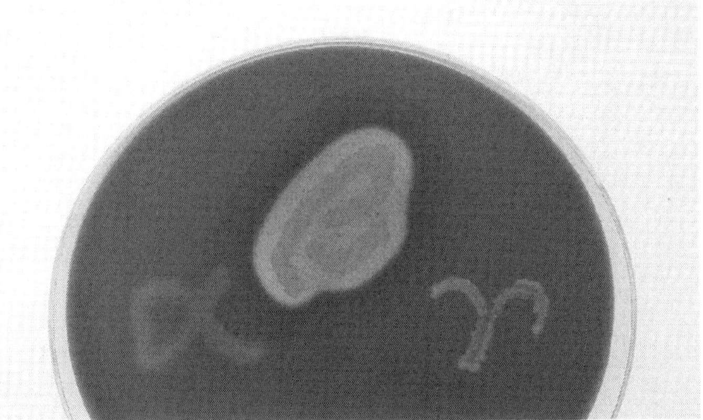

Figure: *Hemolyses of* Streptococcus *spp. (left) á-hemolysis; (middle) â-hemolysis; (right) ã-hemolysis (= non-hemolytic)*

Streptococcus iniae may be easily misidentified (or not identified at all) by conventional automated microbiology systems. Molecular genetics methods, such as DNA sequencing and DNA-DNA hybridization, can be useful for correct identification, although work by the U.S. Centers for Disease Control and Prevention has suggested they are unnecessary in most cases. Several groups have used 16S rDNA sequencing to identify *S. iniae* isolates, and while it can differentiate this species from other related species, such as *S. porcinus* and *S. uberis*, 16S sequencing cannot be used to differentiate between strains of *S. iniae*. Ribotyping is a similar method, by which 16S and 23S rRNA genes are digested with restriction endonucleases and Southern blotted using species-specific oligonucleotide probes. This method is more sensitive than 16S rDNA sequencing, as in addition to species differentiation, it can be used to differentiate between strains. Ribotyping was used in 1997 to differentiate between Israeli and American strains, thus ruling out the possibility of an epidemiological link between outbreaks in the two countries.

Streptococcus iniae is beta-hemolytic when incubated in anaerobic conditions, although it may be misidentified as alpha-hemolytic because, in some strains, zones of beta-hemolysis (complete destruction of red blood cells in the blood agar culture media) are surrounded by large zones of alpha-hemolysis (incomplete destruction of red blood cells with a greenish discoloration due to breakdown of hemoglobin). The bacterium is catalase-negative and LAP-positive (like all streptococci), PYR-test and CAMP-test-positive, does not hydrolyze sodium hippurate and does not grow in bile esculin agar. It does not express any of the known Lancefield antigens.

Serotypes

There are two established serotypes of *S. iniae*. The ATCC 29178 type strain first characterized in 1976 by Pier and Madin is representative of Serotype I isolates. Serotype II was first identified as the type strain (ATCC 29177) isolated from another dolphin case of “golf ball disease”. A biochemical assay measuring arginine dihydrolase activity has been used to distinguish between serotypes (Serotype I is positive), though proposed hyper-encapsulation of Serotype II may represent the most significant functional difference between the two types.

Role in Disease

In Fish: *Streptococcus iniae* is highly pathogenic in freshwater, marine, and euryhaline fish, and is highly lethal: outbreaks may be associated with 30–50% mortality. It is therefore one of the foremost economically important pathogens in intensive aquaculture. In 1997, the global economic impact of *S. iniae* infection to the aquaculture industry was estimated at US$100 million (one-tenth of which in the

United States). As of 2007, infection had been reported in twenty-seven species of fish, including tilapia (genus *Oreochromis* and *Tilapia*), rainbow trout (*Oncorhynchus mykiss*), coho salmon (*Oncorhynchus kisutch*), Japanese amberjack (*Seriola quinqueradiata*), red drum (*Sciaenops ocellatus*), and barramundi (*Lates calcarifer*, which can be an asymptomatic carrier). Common carp (*Cyprinus carpio*), channel catfish (*Ictalurus punctatus*), and goldfish (*Carassius auratus*) appear to be resistant. Fish raised in intensive aquaculture operations and subject to environmental stressors (i.e. suboptimal temperature, poor water quality, crowding, handling, etc.) are most prone to *S. iniae* infection. Wild fish populations located both near and far from aquaculture operations have also proven susceptible to *S. iniae* infection.

Figure: *Epizootics of* S. iniae *infection in rainbow trout (*healthy specimen pictured*) have occurred in Israel and Japan.*

The site of *S. iniae* infection and its clinical presentation vary from species to species. In tilapia, *S. iniae* causes meningoencephalitis, with symptoms including lethargy, dorsal rigidity, and erratic swimming behavior; death follows in a matter of days. In rainbow trout, it is typically associated with septicemia and central nervous system damage. Symptoms are consistent with septicemia, and include lethargy and loss of orientation (as in tilapia), exophthalmia, corneal opacity, and external and internal bleeding.

In Humans: *Streptococcus iniae* can cause opportunistic infections in weakened or immunocompromised humans. It is most commonly associated with bacteremic cellulitis, but has been known to cause endocarditis, meningitis, osteomyelitis, and septic arthritis. The first recognized cases of human infection occurred in Texas in 1991 and in Ottawa in 1994, but the sources of infection were not determined.

Human infection with *S. iniae* was also identified in Toronto between 15 and 20 December 1995, when three Asian patients were admitted to a hospital with cellulitis of the hand after injuring themselves while handling raw fish. All three were found to have bacteremia, initially attributed to *Streptococcus uberis*, but later correctly identified as *S. iniae*.

In February 1996, a Chinese man was admitted to the same hospital with sepsis one week after preparing a fresh whole tilapia, and was also diagnosed with *S. iniae* bacteremia. A subsequent epidemiological investigation found other cases in the Toronto area, for a total of nine patients; all were of Asian descent and all had handled raw fish (mostly tilapia) before developing infection. Other cases were later identified in the United States and elsewhere in Canada, and have since been reported in Asia (Hong Kong, Taiwan, and Singapore). Asian descent is a common trend in the majority of invasive human cases, but it is unknown whether this is due to inherent differences in immunity or because of cultural differences in the fish preparation which lend themselves to a higher incidence of infection.

Control and Treatment

Several measures can be taken to control infection in aquaculture once a *S. iniae* outbreak has been confirmed. Decreasing the quantity of feed given to fish has been shown to reduce mortality rates, as the uptake of bacteria in water is expedited by feeding. Decreasing the density of the fish stock increases survival by reducing injury to fish and lowering the general stress level in the population. Lowering the water temperature and keeping optimal oxygen levels has also been shown to reduce stress to fish and inhibit bacterial growth.

A 2005 study showed the potential for using probiotics for controlling *S. iniae* infection in trout. This study used the gastrointestinal contents of rainbow trout to scan for bacteria that inhibited growth of *S. iniae* and *Lactococcus garvieae*. They identified *Aeromonas sobria* as a potential candidate for control of *S. iniae* and *L. garvieae* infections in aquaculture. *A. sobria*, given live in the feed, protected the trout when challenged with *S. iniae* or *L. garvieae*.

Several antibiotics have been used successfully to treat *S. iniae* infection in fish. Enrofloxacin, a quinolone antibiotic, has been used to great effect in hybrid striped bass (*Morone chrysops* × *M. saxatilis*), although evidence suggested the development of a resistant strain. Amoxicillin, erythromycin, furazolidone, and oxytetracycline have also been used (the last with varying success, only in barramundi). Vaccination against *S. iniae* has been attempted with limited success as it only provides up to 6 months' immunity.

Penicillin has been suggested as the drug of choice for the treatment of *S. iniae* infection in mammals, including humans. In the 1995–1996 cluster of human cases, all clinical isolates were susceptible to penicillin, several cephalosporins, clindamycin, erythromycin, and co-trimoxazole (MICs 0.25 µg/mL); all nine patients were treated with parenteral beta-lactam antibiotics and recovered uneventfully. A study of isolates submitted to the Centers for Disease Control and Prevention between 2000 and 2004 found all to be sensitive to beta-lactams, macrolides, quinolones, and vancomycin.

Taura Syndrome

Taura syndrome is one of the more devastating diseases affecting the shrimp farming industry worldwide.Taura syndrome was first described in Ecuador during the summer of 1992. In March 1993, it returned as a major epidemic and was the object of extensive media coverage. Retrospective studies have suggested that a case of Taura syndrome might have occurred on a shrimp farm in Colombia as early as 1990 and that the virus was already present in Ecuador in mid-1991. Between 1992 and 1997 the disease spread to all major regions of the Americas where *Penaeus vannamei* is cultured. It is estimated that the economic impact of TS in the Americas during that period might have exceeded 2 billion US dollars.

Overview

The 1992 Ecuadorian TS epidemic occurred concurrently with an outbreak of black leaf wilt disease in banana plantations. The outbreak of black leaf disease led to an increase in fungicide usage within the Taura River basin district near the city of Guayaquil. It was thought that the fungicides Tilt (propiconazole, Ciba-Geigy) and Calixin (tridemorph, BASF) used to control black leaf, ran off into nearby ponds and were responsible for the disease. Analytical data demonstrated propiconazole in water, sediments and hepatopancreas tissues of shrimp harvested from affected farms in Ecuador. No other pesticides were discovered.

In January 1994, at the request of Ciba-Geigy, a workshop on Taura syndrome was held at the Aquaculture Pathology Laboratory of the University of Arizona. Experts from several countries with expertise in shrimp and insect pathology, shrimp nutrition, toxicology, myocology, water quality and farm management participated in the workshop. Industry representatives also participated. The group developed recommendations as to the standardization of the research on Taura syndrome and suggested that studies be done to evaluate whether fungicide or an as-yet unrecognized agent was responsible for the syndrome.

Dr. Jim Brock, the aquatic disease specialist for the State of Hawaii during this period,first demonstrated the that the disease could be transmitted by feeding Taura victims to healthy shrimp in early 1994. The dying test shrimp were then fed to a new set of shrimp who was dying at the same rate. Rivers' postulates were fulfilled in 1994 by Dr. Ken Hasson and co-researchers at the University of Arizona. This proved the viral etiology of the syndrome. The virus was named Taura syndrome virus often referred to as TSV. The virus is referred by the name of Infectious cuticular epithelial necrosis virus (ICENV) by some authors in Latin America. Taura syndrome is a notifiable disease by the Office international des Épizooties (OIE), which reflects the serious nature and devastating impact of the disease.

Identification and Description of the Virus

Taura syndrome virus was first classified as a possible member of the family Picornaviridae based on biological and physical characteristics. It was later reclassified in the Dicistroviridae family, genus Cripavirus. It has since been reassigned to a second genus in the same family - the Aparavirus.

TSV is a 32 nm non-enveloped particle with an icosahedral morphology and a buoyant density of 1.338g/ml. The genome is single-stranded positive-sense and has 10,205 nucleotides (excluding the 3' poly-A tail). The capsid consists of three major proteins: CP1 (40 kDa), CP2 (55 kDa) and CP3 (24 kDa) alongside a minor protein of 58 kDa.

Audelo-del-Valle in 2003 reported that certain primate cell lines could be used to culture TSV. Later studies demonstrated that their report was based on misinterpretated data. TSV does not appear to be a potential zoonosis. All virus amplifications require the use of live shrimp, as there is no continuous cell line that supports the growth of shrimp viruses.

Variants of the Virus

RNA viruses such as TSV have high rates of spontaneous mutation. These very high rates might be due to the lack of proofreading function of the RNA-dependent RNA polymerase and have resulted in the emergence of several genetic variants of the virus. As of May 2009, four genetic clusters are recognized: Belize (TSV-BZ), America (TSV-HI), South-east Asia and Venezuela. The Belize strain is considered the most virulent. Point mutations in TSV capsid proteins might provide specific isolate with selective advantages such as host adaptability, increased virulence or increased replication ability. Even small variations in the TSV genome can result in substantial differences in virulence.

All TSV variants are similar in shape and size with light variations. The average size of TSV-BZ virus particles is 32.693+/- 1.834 nm compared to TSV-HI with a size of 31.485 +/- 1.187 nm. The region of highest genetic difference is within the capsid protein CP2 with pairwise comparison of nucleotide showing a 0 to 3.5% difference amongst isolates. The most variations in CP2 occur at the 3'-terminal sequence, this may be because it is less constrained by structural requirements and more exposed than other regions of the protein.

Geographic Distribution

TSV has been reported from virtually all shrimp-growing regions of the Americas including Ecuador, Colombia, Peru, Brazil, El Salvador, Guatemala, Honduras, Belize, Mexico, Nicaragua, Panama, Costa Rica, Venezuela as well as from the States of Hawaii, Texas, Florida and South Carolina. Until 1998, it was considered to be a Western Hemisphere virus. The first Asian outbreak occurred in Taiwan. It has more recently been identified in Thailand, Myanmar, China, Korea and Indonesia where it has been associated with severe epizootics in farmed *Penaeus vannamei* and *Penaeus monodon.*

The wide distribution of the disease has been attributed to the movement of infected host stocks for aquaculture purposes. This might have been helped by the highly stable nature of the virus. It is thought that importation of TSV-infected *Penaeus vannamei* from the Western Hemisphere was at the origin of the outbreak in Taiwan. This was further suggested by the genomic similarity of the Taiwan and Western Hemisphere isolates. TSV appeared in Thailand in 2003. Due to the similarities in deduced CP2 amino acids sequence and the chronology of the disease outbreaks in relation to imported stocks it is likely that at least some of the Thai isolates originated from Chinese stocks.

Taura syndrome can spread rapidly when introduced in news areas. A shrimp farmer described the 1995 outbreak in Texas as "This thing spread like a forest fire... There was no stopping it. I just sat there and watched it, and in a matter of three days, my shrimp were gone. Dead!" (reported in Rosenberry: Shrimp news international in 1995).

Species of Susceptible Shrimp

TSV is known to affect many shrimp species. It causes serious diseases in the postlarval (PL), juvenile and adult stages of *Penaeus vannamei*. It also affects severely *Penaeus setiferus, Penaeus stylirostris, Penaeus schmitti, Metapenaeus ensis. Penaeus chinensis* are highly susceptible to the disease in experimental bioassays.

There is variation within species and TSV-resistant strains of shrimp have been developed. Wild stocks are showing increased resistance, perhaps through intense natural selection. Reports of Taura syndrome in the wild are limited, but in February 1995, the Mexican Fisheries Ministry reported the presence of TSV in wild-type shrimp captured on the border of Mexico and Guatemala. As of 2007, there were no confirmed reports indicating that TSV is infectious to other groups of decapod or non-decapod crustaceans.

Pathology and Disease Cycle

In farm situations, TS often causes high mortality during the first 15 to 40 days of stocking into shrimp ponds. The course of infection may be acute (5–20 days) to chronic (more than 120 days) at the pond and farm level. The disease has three distinct phases that sometimes overlap: acute, transition and chronic. The disease cycle has been characterized in detail in *Penaeus vannamei*.

After the initial infection, the acute phase develops. Clinical signs can occur as early as 7 hours post-infection in some individuals and last for about 4–7 days. Infected shrimp display anorexia, lethargy and erratic swimming behavior. They also present opacification of the tail musculature, soft cuticle and, in naturally occurring infection, a red tail due to the expansion of the red chromatophores. Mortality during this phase can be as high as 95%. The acute phase is characterized histologically by multifocal areas of nuclear pyknosis/ karyorrhexis and numerous cytoplasmic inclusion bodies in the cuticular epithelium and the subcutis of the general body surface, all appendages, gills, hindgut, esophagus and stomach. The pyknosis and karyorrhexis give a "buckshot" appearance to the tissue and are considered pathognomonic for the disease. In severe infections the antennal gland tubule epithelium, the hematopoietic tissues and the testis are also affected. This occurs mainly in severe infection following injection of viral particles and has not been reported from naturally infected *Penaeus vannamei*. Shrimp that survive the acute stage enter a transitional stage.

Shrimp in the transitional phase show randomly distributed melanized (brownish/black) lesions within of the cuticle of the cepahlothorax and tail region. These foci are the sites of acute lesions which have progressed onto subsequent stages of hemocytic inflammation, cuticular epithelium regeneration and healing, and which might be secondarily infected with bacteria. These foci are negative for TSV by in situ hybridization (ISH) using a TSV specific cDNA probe. Histologically these shrimp present focal active acute lesions and the onset of lymphoid organ spheroids (LOS) development. By ISH with TSV specific probes a diffuse positive signal can be observed within the walls of the lymphoid organ of normal appearance with or without focal probe signals within developing LOS. These shrimp will be lethargic and anorexic, possibly because of the redirection of all their energy and metabolic resources toward wound repair and recovery. If the shrimp undergo another successful moult following the transitional phase they will cast off the melanized lesions and enter the chronic phase.

The chronic phase is first seen 6 days post-infection and persist for at least 12 months under experimental conditions. This phase is characterized histologically by the absence of acute lesions and the presence of LOS of successive morphologies. These LOS are positive by ISH for TSV. A low prevalence of ectopic spheroids can also be observed in some cases. LOS are not by themselves characteristic of TSV infection and can be found in other viral diseases of shrimp such as lymphoid organ vacuolization virus (LOVV), lymphoid parvo-like virus (LPV), lymphoid organ virus (LOV), rhabdovirus of penaeid shrimp (RPS) and yellow head virus (YHV). Diagnosis of the disease during the chronic phase is problematic as shrimp do not display any outward signs of the disease and do not show mortality from the infection. Survivors may become carrier for life. Shrimp with chronic TSV infection are as vigorous as uninfected shrimp, as demonstrated by their inability to tolerate a salinity drop as well as uninfected shrimp.

Routes of Transmission

The most likely route for transmission of TSV is cannibalism of dead infected shrimp. The virus can be spread from one farm to another by seagulls and aquatic insects. Infectious TSV has been found in the feces of laughing gulls (*Larus atricilla*) that fed on infected shrimp during an epizootic in Texas. Controlled laboratory studies have documented that TSV remains infectious for up to one day after passage through the gut of white leghorns chicken (*Gallus domesticus*) and laughing gulls. Although, vertical transmission is suspected this has not been experimentally confirmed.

Shrimp surviving a TSV infection are life-long carriers of the viruse and are a significant source of virus for susceptible animals. It has been hypothesized that TSV was introduced to Southeast Asia with chronically infected shrimp imported from the Western Hemisphere. The ability of TSV to remain at least partly infectious after one or several freeze-thaw cycles might be a contributing factor facilitating its spread in the international commerce of frozen commodity products. Mechanisms by which infected frozen shrimp could spread the virus include: reprocessing of shrimp at processing plants with release of infectious liquid wastes, disposal of solid wastes in landfills where seagulls could acquire the virus and then spread it, the use of shrimp as bait by sport fishermen and the use of imported shrimp as fresh food for other aquatic species.

Diagnostic Methods

A presumptive diagnosis of acute TSV infection can be established by the presence of dead or dying shrimp in cast nets used for routine evaluation. Predatory birds are attracted to diseased ponds and feed heavily on the dying shrimp. The unique signs of infection caused by TS such as the cuticular melanized spots can provide a strong presumptive diagnosis, but care must be taken as these can be confused with other diseases, such as bacterial shell disease. In general pathognomonic histopathological lesions are the first step in confirmatory diagnosis. Discrete foci of pyknotic and karyorhectic nuclei and inflammation are seen within the cuticular tissues. The lymphoid organ might display spheroids, but is otherwise unremarkable.

The genome of the virus has been cloned and cDNA probes are available for diagnosis. Reverse transcriptase polymerase chain reaction (RT-PCR) methods have been developed for detection of TSV and are very sensitive. Real-time techniques allow for quantification of the virus. The IQ2000TM TSV detection system, a RT-PCR method, is said to have a detection limit of 10 copies per reaction.

RNA-based methods are limited by the relative fragility of the viral RNA. Prolonged fixation in Davidsons' fixative might result in RNA degradation due to fixative-induced acid hydrolysis.

An alternative for virus detection is the utilization of specific monoclonal antibodies (MAbs) directed against the relatively stable proteins in the viral capsid. Rapid diagnostic tests using MAbs are now in common use for white spot syndrome virus and are being marketed under the commercial name of *Shrimple*. Similar tests for TSV, Yellowhead virus and Infectious Hypodermal and Haematopoietic Necrosis Virus are currently under development.

Methods of Control

Management strategies for the disease have included raising more resistant species such as the Western blue shrimp (*Penaeus stylirostris*) and stocking of specific pathogen free (SPF) or specific pathogen resistant (SPR) shrimp.

Relatively simple laboratory challenges can be used to predict the performance of selected stocks in farm where TSV is enzootic. The resistant shrimp lines currently raised have reached nearly complete resistance to some TSV variants and further improvement due to breeding for TSV-resistance is expected be minor for these variants. Significant improvements in TSV survival were made through selective breeding despite low to moderate heritability for this trait.

A management strategy used to reduce the impact of TS has been the practice of stocking post larva shrimp at increased stocking density. Following this strategy, farms would experience mortality due to TS at an early stage of the production cycle, before substantial feeding had begun, and the surviving shrimp would be resistant to further TSV challenges.

Other techniques used with limited efficacy have been the polyculture of shrimp with tilapia and maintenance of near optimal water quality conditions in the grow-out ponds with reduction of organic loading. Transgenic shrimp expressing an antisense TSV coat protein (TSV-CP) exhibited increased survival in TSV challenges. The public perception of transgenic animals as well as current technical limitations, limit the use of transgenic animals as a mean of disease control.

Ulcerative Dermal Necrosis

Ulcerative dermal necrosis (UDN) is a chronic dermatological disease of cold water salmonid fish that had a severe impact on north Atlantic Salmon and sea trout stocks in the late 1960s, the 1970s and 1980.

Affected fish developed severe skin lesions over large parts of their body which penetrated into skeletal muscle.

The onset of symptoms only occurred after migration into freshwater. Lesions became quickly infected with overgrowths of *Saprolegnia* fungus giving the affected fish an appearance of being covered in slimy white pustules. The most severely affected fish frequently die before spawning.

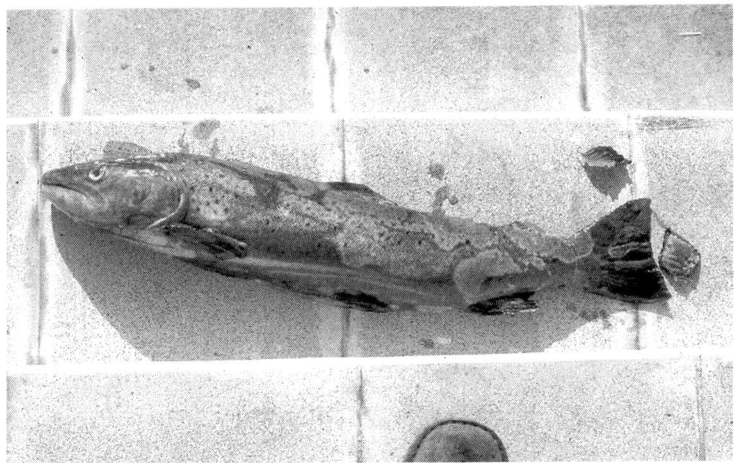

Figure: *Sea trout affected by UDN with typical secondary Saprolegnia infections*

Although the worst effects of the disease were seen in the 1970s and 1980, even now large numbers of salmon will succumb to the disease after spawning. This is thought be due in part to their weak post-spawning condition, and lack of food for several months whilst in the river.

Those fish that do make it back to the sea are thought to make a good recovery.

Viral Hemorrhagic Septicemia

Viral hemorrhagic septicemia (VHS) is a deadly infectious fish disease caused by the *Viral hemorrhagic septicemia virus* (VHSV, or VHSv). It afflicts over 50 species of freshwater and marine fish in several parts of the northern hemisphere. VHS is caused by the viral hemorrhagic septicemia virus (VHSV), different strains of which occur in different regions, and affect different species. There are no signs that the disease affects human health. VHS is also known as "Egtved disease," and VHSV as "Egtved virus."

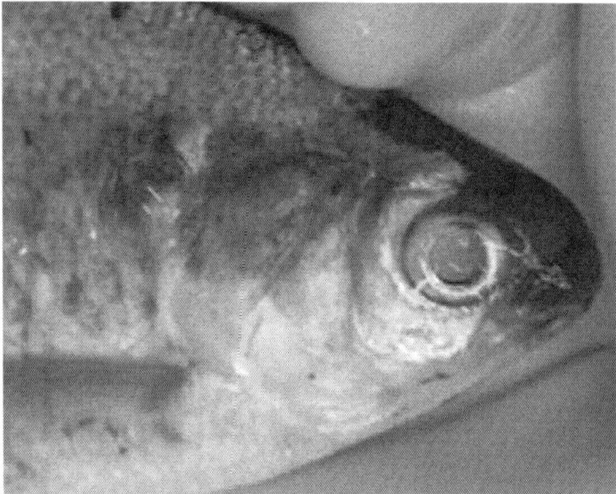

Figure: *VHS disease in a gizzard shad*

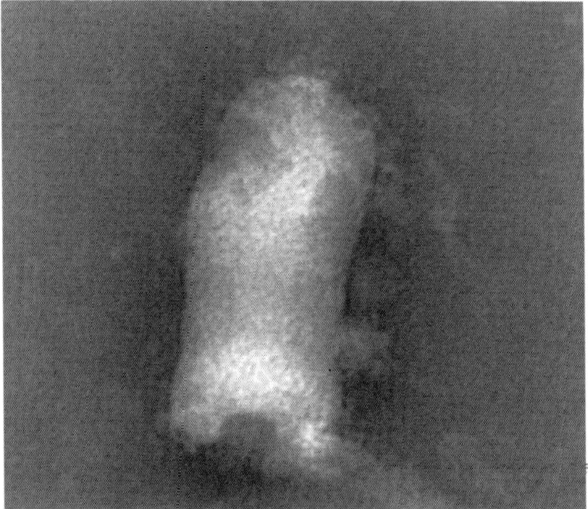

Figure: *Electron Micrograph of VHSV*

Historically, VHS was associated mostly with freshwater salmonids in western Europe, documented as a pathogenic disease among cultured salmonids since the 1950s. Today it is still a major concern for many fish farms in Europe and is therefore being watched closely by the European Community Reference Laboratory for Fish Diseases. It was first discovered in the US in 1988 among salmon returning from the Pacific in Washington State. This North American genotype was identified as a distinct, more marine-stable strain than the European genotype. VHS has since been found afflicting marine fish in the northeastern Pacific Ocean, the North Sea, and the Baltic Sea. Since 2005, massive die-offs have occurred amongst a wide variety of freshwater species in the Great Lakes region of North America.

Virus Taxonomy

VHSV is a negative-sense single-stranded RNA virus of the order *Mononegavirales*, family *Rhabdoviridae*, and genus *Novirhabdovirus*. (ICTV virus classification, unlike biological classification, begins with the order.) Its NCBI-assigned Taxonomy ID is 11287. Another related fish rhabdovirus in the *Novirhabdovirus* genus is the Infectious hematopoietic necrosis virus (IHNV), which causes infectious hematopoietic necrosis (IHN) disease in solmonidae.

The viral cause of the disease was discovered in 1963 by M. H. Jenson. The virus is an enveloped, bullet-shaped particle, about 180 nm long by 60 nm in diameter, covered with 5 to 15 nm long peplomers.

Molecular Virology

The genome of VHSV is composed of approximately 11-kb of single stranded RNA, which contains six genes that are located along the genome in the 32 -52 order: 32 -N-P-M-G-NV-L-52 , nucleocapsid protein (N), polymerase-associated phosphoprotein (P), matrix protein (M), surface glycoprotein (G), a unique non-virion protein (NV), and virus polymerase (L).

Reverse genetics is a powerful tool to study and characterize the previously unknown viral genes. Reverse genetics system is currently available for VHSV. A vaccinia virus free reverse genetic system for Great Lakes VHSV (Genotype IVb) was developed by a research group from the USA. This system allows the investigators to explore the functional properties of individual viral genes of VHSV in detail. This system was immediately utilized to characterize the non-virion (NV) gene of *Novirhabdoviruses*. Even though it has been demonstrated that the NV gene is not necessary for viral replication, it is highly essential for viral pathogenesis. A new role of NV protein has been discovered and demonstrated that it inhibits apoptosis at the early stage of viral infection. This discovery unlocked the mystery of presence of NV protein in *Novirhabdoviruses*.

Virus Subtypes

Different isolates (unique strains) of VHSV are typically grouped by genotyping. It is found that genotype groups are divided more geographically than by host species. Earlier studies used different numbering systems, but the following system has come into common usage based on genotype similarity based on sequencing of the N- and G-genes. Types I-III are enzootic to Europe, and Type IV to North America, and Type I and type IV isolates are further subdivided, as follows:

Type	***Prevalent host type and location***
I-a	Farmed rainbow trout and a few other freshwater fish in continental Europe
I-b	Marine fish of the Baltic Sea, Skagerrak, Kattegat, North Sea, Japan
I-c	Farmed rainbow trout Denmark
I-d	Farmed rainbow trout in Norway, Finland, Gulf of Bothnia
I-e	Rainbow trout in Georgia, farmed and wild turbot in the Black Sea
II	Marine fish of the Baltic Sea
III	Marine fish of the British Isles and northern France, farmed turbot in the UK and Ireland, and Greenland halibut (*Reinhardtius hippoglossoides*) in Greenland
IV-a	Marine fish of the Northwest Pacific (North America), North American north Atlantic coast, Japan, and Korea
IV-b	Freshwater fish in North American Great Lakes region

Type I-a was the only strain known from VHSV's discovery in 1963 until the late 1988, isolated to fish farms in continental Europe, affecting primarily rainbow trout and occasionally brown trout or pike.

In 1988, the first marine strain of VHSV, now designated type IV, was found in normal-appearing salmon returning from the Pacific to rivers of Washington State. This strain and other marine strains was not lethal to rainbow trout. The discovery prompted further studies, and by the mid-1990s, marine VHSV was found in eight species along the northern North America's Pacific coast, and 14 species in and around the Atlantic's North Sea. 1996 saw the first VHSV in Japan, among Japanese flounder farmed in the Seto Inland Sea, and different genotypes have appeared in different areas since then. Type IV was later found off North America's northern Atlantic coast, in Atlantic herring (*Clupea harengus*) mummichog (*Fundulus heteroclitus*), stickleback (*Gasterosteus aculeatus aculeatus*), brown trout (*Salmo trutta*), and striped bass (*Morone saxatilis*), as well as dozens of freshwater species in the Great Lakes.

VHSV continues to be found in new geographical areas, in new species of fish. This is thought to represent both the spread of the virus into new areas, as with VHSV egg and live fish transfers from North America to Asia, or feeding of raw marine fish to inland farmed trout in Finland, as well as discovery of existing populations, as with an apparently well established marine reservoir in the Black Sea.

To keep track of the distribution of different VHSV genotypes, a database called Fishpathogens.eu has been created to store data on different fish pathogens (including VHSV) and their sequences.

Great Lakes Type IV-b

The Type IV-b strain of VHSV has been spreading among freshwater fish in the Great Lakes region since at least 2003, resulting in some massive die-offs since 2005 of many species in the affected lakes. Originally found off the Atlantic coast of Canada, it was considered a low mortality marine strain. Its first detection in freshwater was in Lake Ontario in 2005, and then in an archived 2003 sample from Lake St. Clair. The isolate was named MI03GL and was sequenced for its entire genome.

The North American genotype of the virus, in addition to moderate mortality to salmonid species, including various varieties of trout, is also proving virulent among a wide variety of warm-water species previously considered resistant to VHS. The Great Lakes region variant has killed lake trout, steelhead trout, chinook salmon, yellow perch, gobies, emerald shiners, muskies, whitefish, and walleye. While the European strain of VHSV is particularly deadly to rainbow trout, the Great Lakes region variant affects the species only mildly, as is typical with primarily marine genotypes.

Great Lakes Regional Distribution

An archived 2003 sample from Lake St. Clair of Great Lakes muskellunge is the earliest confirmed case of VHSV within the Great Lakes region. Lake St. Clair connects to Lake Erie through the Detroit River to the south, and to Lake Huron through the St. Clair River to the north. The sample was not tested for VHS until 2005, after the disease was detected in Lake Ontario.

2005 samples of Lake Ontario freshwater drum and Lake Huron lake whitefish were infected with VHS. Initially classified as an unknown rhabdovirus, the Lake Ontario sample was confirmed to be VHSV in 2006, while the Lake Huron sample was confirmed to be VHSV in 2007.

2006 saw mass die-offs from VHS in Lake Erie, the St. Clair River, the Detroit River, and the St. Lawrence River, which connects the Great Lakes to the Atlantic Ocean. VHS was further detected in the Niagara River, which connects Lake Erie to Lake Ontario. It was also found in a walleye die-off in the landlocked inland Conesus Lake, the westernmost of the Finger Lakes in western New York state. This was the first case in the region outside waters contiguously connected to the Great Lakes.

On May 12, 2007, the Wisconsin DNR announced the likely presence of VHS in Wisconsin's inland Little Lake Butte des Morts. Preliminary tests of samples of freshwater drum collected on May 2 were positive, and the announcement came amidst a die-off of hundreds of freshwater drum there and on neighboring Lake Winnebago. Preliminary tests later indicated VHS in specimens from Lake Winnebago. The lakes drain through the Fox River to Lake Michigan's Green Bay.

On May 17, 2007, the Michigan DNR confirmed the presence of VHS in the Michigan's inland Budd Lake, a popular fishing destination is in the central part of Michigan's lower peninsula. A major die-off of VHS-positive muskies, bluegills, and black crappie began on April 30, 2007.

On May 24, 2007, preliminary tests indicated the presence of VHS in a brown trout from Lake Michigan, the second largest freshwater lake in the United States. Contamination in the lake had been expected for months by experts, since the presence of VHS was confirmed in the connected Lake Huron.

On July 14, 2007, federal labs confirmed the presence of VHS in Skaneateles Lake, the second of New York's Finger Lakes to test positive for the disease. The disease caused a large die-off of bass in the spring of 2007.

Government Regulation

Before the Type IVb die-offs in the Great Lakes, few states regulated transfer of live species other than for salmonids. Since 2005, new policies have been adopted concerning fish and egg transfer, use of live bait, and water transfer, aimed at curtailing the spread to new lakes and rivers in the region.

As of July 13, 2007, new rules have been enacted in the Canadian province of Ontario, and US states of Michigan, New York, Ohio, Pennsylvania, and Wisconsin, while they are currently being drafted in Illinois, Indiana, and Minnesota. Additionally, the USDA's Animal and Plant Health Inspection Service issued a federal order in the fall of 2006 barring the transfer of all live susceptible species from the eight states bordering the lower Great Lakes, as well as importing such species from The Canadian provinces of Ontario and Quebec.

Transmission

VHSV can be spread from fish to fish through water transfer, as well as through contaminated eggs, and bait fish from infected waters. The emerald shiner is a particularly popular bait fish in the Great Lakes region, and is among the species afflicted.

Survivors of the disease can become lifelong carriers of the virus, contaminating water with urine, sperm, and ovarian fluids. The virus has been shown to survive two freeze/thaw cycles in a conventional freezer, suggesting both live and frozen bait could be a transmission vector. In Europe, the gray heron has spread the virus, but it does so mechanically; the virus is apparently inactive in the digestive tract of birds.

Symptoms

Fish that become infected experience hemorrhaging of their internal organs, skin, and muscle. Some fish show no external symptoms, but others show signs of infection that include bulging eyes, bloated abdomens, bruised-looking reddish tints to the eyes, skin, gills and fins. Some infected fish have open sores that may look like the lesions from other diseases or from lamprey attacks. There may also be a nervous form of the disease where fish are constantly flashing and showing abnormal behaviour.

Diagnosis

Field Diagnosis: Living fish afflicted with VHS may appear listless or limp, hang just beneath the surface, or swim very abnormally, such as constant flashing circling due to the tropism of the virus for the brain.

External signs may include darker coloration, exophthalmia ("pop eye"), pale or red-dotted gills, sunken eyes, and bleeding around orbits (eye sockets) and at base of fins.

Genetics researchers at the Lake Erie Research Center at the University of Toledo are developing a test that will speed diagnosis from a month to a matter of hours.

Gross Pathology (Non-laboratory)

VHSV is a hemorrhagic disease, meaning it causes bleeding. Internally, the virus can cause petechial hemorraging (tiny spots of blood) in internal muscle tissue, and petechial or severe hemorrhaging in internal organs and other tissues. Internal hemorrhaging can be observed as red spots inside a dead fish, particularly around the kidney, spleen, and intestines, as well as the swim bladder, which would normally have a clear membrane. The liver may be pale, mottled with red hyperemic areas, the kidney may be swollen and unusually red, the spleen may be swollen, and the digestive tract may be empty.

External signs are not always present, but if they are, hemorrhaging on the skin's surface can appear as anywhere from tiny red dots (petechiae) to large red patches.

Histopathology (Microscopic tissue analysis)

Preliminary diagnosis involves histopathological examination, observing tissues through a microscope. Most tissue changes can be observed as minor to major necrosis (cell death) in the liver, kidneys, spleen, and skeletal muscle. The hematopoietic (blood-forming) areas of the kidney and spleen are the initial area of infection, and should show necrosis. The gill may have thickened lamellae, and the liver may have pyknotic nuclei. Skeletal muscle accumulates blood but does not suffer much damage.

Virology (Definitive testing)

Electron microscopy can reveal the bullet-shaped rhabdovirus, but is not adequate for definitive diagnosis. The Manual or Diagnostic for Aquatic Animals, 2006, is the standard reference for definitive tests. In most cases, cell culturization is recommended for surveillance, with antibody tests and reverse transcription polymerase chain reaction (RT-PCR) and genetic sequencing and comparison for definitive confirmation and genotype classification. Virus neutralisation is another important method of diagnosis, especially for carrier fish.

Prevention

Thoroughly cleaning boats, trailers, nets and other equipment when traveling between different lakes and streams also helps. The only EPA-approved disinfectant proven effective against VHS is Virkon AQUATIC (made by Dupont). Chlorine bleach kills the VHS virus, but in concentrations that are much too caustic for ordinary use. Disinfecting stations can be found at various inland lake boat launches in the Great Lakes region.

White Spot Syndrome

White spot syndrome (WSS) is a viral infection of penaeid shrimp. The disease is highly lethal and contagious, killing shrimps quickly. Outbreaks of this disease have wiped out within a few days the entire populations of many shrimp farms throughout the world. The disease is caused by a family of related viruses subsumed as the White Spot Syndrome Baculovirus Complex (WSSV) and the disease caused by them as white spot syndrome (WSS).

History

The first reported epidemic due to this virus is from Taiwan in 1992, however, reports of losses due to white spot disease came from China in 1993, where it led to a virtual collapse of the shrimp farming industry. This was followed by outbreaks in Japan and Korea in the same year, Thailand, India and Malaysia in 1994 and by 1996 it had severely affected East Asia and South Asia. In late 1995, it was reported in the USA, 1998 in Central and South America, 1999 in Mexico and in 2000 in the Philippines. Currently, it is known to be present in all shrimp growing regions except Australia.

Virology

WSSV is a rod-shaped double-stranded DNA virus, and the size of the enveloped viral particles have been reported to be 240–380 nm long and 70–159 nm in diameter and nucleocapsid core is 120–205 nm long and 95–165 nm in diameter. The virus has an outer lipid bilayer membrane envelope, sometimes with a tail like appendage at one end of the virion. The nucleocapsid consists of 15 conspicuous vertical helices located along the long axis, each helix has two parallel striations, composed of 14 globular capsomers, each of which is 8 nm in diameter.

The complete DNA sequence of WSSV genome has been assembled into a circular sequence of 292,967 bp. It encodes 531 putative open reading frames.

One of the proteins – WSSV449 – has some similarity to host protein Tube and can function like Tube by activating the NF-êB pathway.

Transmission of the virus is mainly through oral ingestion and water borne routes in farms (horizontal transmission) and vertical transmission (from infected mother prawns) in case of shrimp hatcheries. The virus is present in the wild stocks of shrimp, especially in the coastal waters adjacent to shrimp farming regions in Asian countries, but mass mortalities of wild shrimps are yet to be observed.

Clinical

The virus has a wide host range and is highly virulent and leads to mortality rates of 100% within days in the case of cultured penaeid shrimps. Most of the cultured penaeid shrimps (*Penaeus monodon,*

Marsupenaeus japonicus, Litopenaeus vannamei, Fenneropenaeus indicus) are natural hosts of the virus. Several non-penaeid shrimps were also found to be severely infected during experimental challenges. Many crustaceans like crabs (*Scylla spp., Portunus spp.*), spiny lobsters (*Panulirus spp.*), crayfish (*Astacus spp., Cherax spp.*) and freshwater shrimp (*Macrobrachium spp.*) are reported to be infected with variable severities depending on the life stage of the host and presence of external stressors (temperature, salinity, bacterial diseases, pollutants).

Clinical signs of WSSV include a sudden reduction in food consumption, lethargy, loose cuticle and often reddish discolouration, and the presence of white spots of 0.5 to 2.0 mm in diameter on the inside surface of the carapace, appendages and cuticle over the abdominal segments.

Pathology

In the host WSSV infects a wide variety of cells from ectodermal and mesodermal origin. Histological changes are seen in the gill epithelium, antennal gland, haematopoeitic tissue, nervous tissue, connective tissue and intestinal epithelial tissue. Infected cells have prominent intranuclear occlusions that initially stain eosinophilic but become basophilic with age; hypertrophied nuclei with chromatin margination; and cytoplasmic clearing. Pathogenesis involves widespread tissue necrosis and disintegration.

White spots on the shell of infected shrimp under scanning electron microscope appear as large dome shaped spots on the carapace measuring 0.3 to 3 mm in diameter. Smaller white spots of 0.02 to 0.1 mm appear as linked spheres on the cuticle surface. Chemical composition of the spots is similar to the carapace, calcium forming 80–90% of the total material and it is suggested to have derived from abnormalities of the cuticular epidermis.

A number of biochemical changes have been reported after infection with this virus: glucose consumption and plasma lactate concentration increase, glucose 6 phosphate dehydrogenase activity increases and triglyceride concentration decreases. The voltage dependent anion channel of the mitochondrion is also up regulated.

Diagnosis

Infection with WSSV differs from other described penaeid infections Yellowhead virus (YHV) and Infectious Hypodermal and Hematopoietic Necrosis virus (IHHNV) in the described histological findings as YHV has a reduced tissue specificity, infecting only the intestinal epithelial tissues and IHHNV causes intranuclear occlusions that stain eosinophillic but do not change over the course of the infection. Rapid and specific diagnosis of the virus can be accomplished using nested or real-time polymerase chain reaction (PCR).

Treatment

There are no available treatments for WSS.

Prevention

A large number of disinfectants are widely used in shrimp farms and hatcheries to prevent an outbreak. Stocking of uninfected shrimp seeds and rearing them away from environmental stressors with extreme care to prevent contamination are useful management measures.

Yellowhead Disease

Yellowhead disease (YHD) is a viral infection of shrimp and prawn, in particular of the giant tiger prawn (*Penaeus monodon*), one of the two major species of farmed shrimp. The disease is highly lethal and contagious, killing shrimp quickly. Outbreaks of this disease have wiped out in a matter of days the entire populations of many shrimp farms that cultivated *P. monodon*, i.e. particularly Southeast Asian farms. In Thai, the disease is called *Hua leung*. The disease is caused by the yellowhead virus (YHV), a positive-sense single-stranded RNA virus related to coronaviruses and arteriviruses. A closely related virus is the gill-associated virus (GAV), which is the type species of the genus *Okavirus*.

Clinical

The cephalothorax of infected shrimp turns yellow after a period of unusually high feeding activity ending abruptly, and the then moribund shrimps congregate near the surface of their pond before dying. YHD leads to death of the animals within two to four days.

History

YHD had been reported first from Thailand in 1990, the closely related GAV has been discovered in 1995 during a yellowhead-like disease in Australian shrimp farms.

Amoebic Gill Disease

Amoebic gill disease (AGD) is a potentially fatal disease of some marine fish.

It is caused by *Neoparamoeba pemaguidensis* and *Neoparamoeba perurans*, some of the most important amoebas in cultured fish. It primarily affects farm raised fish of the Salmonidae family, most notably affecting the Tasmanian Atlantic Salmon (Salmo salar) industry, costing the AU$230 million a year in treatments and lost productivity. Turbot, bass, bream, sea urchins and crabs have also been infected. The disease has also been reported affecting the commercial salmon fisheries of the United States, Australia, New Zealand, France, Spain, Ireland and Chile. It was first diagnosed in the summer of 1984/1985 in populations of Atlantic Salmon off the east coast of Tasmania and was found to be caused by the *Neoparamoeba perurans* n.sp.

Clinical Signs and Diagnosis

Symptoms typically begin to appear two months after the fish are transferred from freshwater hatcheries to open net sea cages. Symptoms include mucus build-up on the gills of infected fish and hyper-plastic lesions, causing white spots and eventual deterioration of the gill tissue. Fish will show signs of dyspnoea such as rapid opercular movements and lethargy. Contributing factors are an ambient water temperature above 16 degrees Celsius, crowding and poor water circulation inside the sea pens. Clinical cases are more common in the Summer. The lesions on the gills are highly suggestive of infection. Gill biopsies can be observed under the mciroscope for amoebas, or tested using fluorescent antibody testing.

Treatment and Control

Currently, the most effective treatment is transferring the affected fish to a freshwater bath for a period of 2 to 3 hours. This is achieved by towing the sea cages into fresh water, or pumping the fish from the sea cage to a tarp filled with fresh water.

Mortality rates have been lowered by adding Levamisole to the water until the saturation is above 10ppm. Due to the difficulty and expense of treatment, the productivity of salmon aquaculture is limited by access to a source of fresh water. Chloramine and chlorine dioxide have also been used.

Argulidae

The family Argulidae contains the carp lice or fish lice – a group of parasitic crustaceans of uncertain position within the Maxillopoda. Although they are thought to be primitive forms, they have no fossil record. Argulidae is the only family in the order Arguloida (occasionally

"Arguloidea"), although a second family, Dipteropeltidae, has been proposed.

Description

Fish lice vary in size from just a few millimetres to over 30 millimetres (1.2 in) long, with females usually somewhat larger than the males. Almost all species in the family are ectoparasites on fish, with a few on invertebrates or amphibians. They have a flattened, oval body, which is almost entirely covered by a wide carapace. Their compound eyes are prominent, and the mouthparts and the first pair of antennae are modified to form a hooked, spiny proboscis armed with suckers, as an adaptation to parasitic life. They have four pairs of thoracic appendages, which are used to swim when not attached to the host. They leave their hosts for up to three weeks in order to mate and lay eggs, and reattach behind the fish's operculum, where they feed on mucus and sloughed-off scales, or pierce the skin and feed on the internal fluids. The eggs hatch into parasitic postnauplius larvae.

Classification

There are 173 recognised species, divided among six genera. The centres of diversity are the Afrotropical and Neotropical ecozones.

Argulus O. F. Müller, 1785:

- *Argulus africanus* Thiele, 1900
- *Argulus alexandrensis* C. B. Wilson, 1923
- *Argulus alosae* Gould, 1841
- *Argulus amazonicus* Malta & Santos-Silva, 1986
- *Argulus ambloplites* C. B. Wilson, 1920
- *Argulus ambystoma* Poly, 2003
- *Argulus americanus* C. B. Wilson, 1902
- *Argulus angusticeps* Cunnington, 1913
- *Argulus annae* Schuurmans Stekhoven J.H. Jr, 1951
- *Argulus appendiculosus* C. B. Wilson, 1907
- *Argulus arcassonensis* Cuénot, 1912
- *Argulus argulus* Leach, 1814
- *Argulus armiger* O. F. Müller, 1785
- *Argulus australiensis* Byrnes, 1985
- *Argulus belones* Kampen, 1909

- *Argulus bengalensis* Ramakrishna, 1951
- *Argulus bicolor* Bere, 1936
- *Argulus biramosus* Bere, 1931
- *Argulus boli* Tripathi, 1975
- *Argulus borealis* C. B. Wilson, 1912
- *Argulus brachypeltis* Fryer, 1959
- *Argulus caecus* C. B. Wilson, 1922
- *Argulus canadensis* C. B. Wilson, 1916
- *Argulus capensis* Barnard, 1955
- *Argulus carteri* Cunnington, 1931
- *Argulus catostomi* Dana & Herrick, 1837
- *Argulus cauveriensis* Thomas & Devaraj, 1975
- *Argulus charon* O. F. Müller, 1785
- *Argulus chesapeakensis* Cressey, 1971
- *Argulus chicomendesi* Malta & Varella, 2000
- *Argulus chilensis* Martinez, 1952
- *Argulus chinensis* Ku & Yang, 1955
- *Argulus chromidis* Kroyer, 1863
- *Argulus confuscus* Rushton-Mellor, 1994
- *Argulus coregoni* Thorell, 1865
- *Argulus cubensis* C. B. Wilson, 1936
- *Argulus dactylopteri* Thorell, 1865
- *Argulus dageti* Dollfus, 1960
- *Argulus dartevellei* Brian, 1940
- *Argulus delphinus* O. F. Müller, 1785
- *Argulus diversicolor* Byrnes, 1985
- *Argulus diversus* C. B. Wilson, 1944
- *Argulus ellipticaudatus* K. N. Wang, 1960
- *Argulus elongatus* Heller, 1857
- *Argulus ernsti* Weibezahn & Cobo, 1964
- *Argulus exiguus* Cunnington, 1913
- *Argulus flavescens* C. B. Wilson, 1916
- *Argulus floridensis* Meehan, 1940
- *Argulus fluviatilis* Thomas & Devaraj, 1975

- *Argulus foliaceus* (Linnaeus, 1758)
- *Argulus fryeri* Rushton-Mellor, 1994
- *Argulus funduli* Krøyer, 1863
- *Argulus fuscus* Bere, 1936
- *Argulus giordanii* Brian, 1959
- *Argulus gracilis* Rushton-Mellor, 1994
- *Argulus hylae* Lemos de Castro & Gomes-Correa, 1985
- *Argulus ichesi* Bouvier, 1910
- *Argulus incisus* Cunnington, 1913
- *Argulus indicus* Weber, 1892
- *Argulus ingens* C. B. Wilson, 1912
- *Argulus intectus* C. B. Wilson, 1944
- *Argulus izintwala* J. G. Van As & L. L. Van As, 2001
- *Argulus japonicus* Thiele, 1900
- *Argulus jollymani* Fryer, 1956
- *Argulus juparanensis* Lemos de Castro, 1950
- *Argulus kosus* Avenant-Oldewage, 1994
- *Argulus kunmingensis* Shen, 1948
- *Argulus kusafugu* Yamaguti & Yamasu, 1959
- *Argulus laticauda* S. I. Smith, 1874
- *Argulus latus* S. I. Smith, 1874
- *Argulus lepidostei* Kellicott, 1877
- *Argulus longicaudatus* C. B. Wilson, 1944
- *Argulus lunatus* C. B. Wilson, 1944
- *Argulus macropterus* Heegaard, 1962
- *Argulus maculosus* C. B. Wilson, 1902
- *Argulus major* K. N. Wang, 1960
- *Argulus mangalorensis* Natarajan, 1982
- *Argulus matuii* Sikama, 1938
- *Argulus meehani* Cressey, 1971
- *Argulus megalops* S. I. Smith, 1874
- *Argulus melanostictus* C. B. Wilson, 1935
- *Argulus melita* Beneden, 1891
- *Argulus mexicanus* Pineda, Paramo & del Rio, 1995

- *Argulus mississippiensis* C. B. Wilson, 1916
- *Argulus mongolianus* Tokioka, 1939
- *Argulus monodi* Fryer, 1959
- *Argulus multicolor* Schuurmans Stekhoven J.H. Jr, 1937
- *Argulus multipocula* Barnard, 1955
- *Argulus nativus* Kirtisinghe, 1959
- *Argulus natterei* Heller, 1857
- *Argulus niger* C. B. Wilson, 1902
- *Argulus nobilis* Thiele, 1904
- *Argulus onodai* Tokioka, 1936
- *Argulus papuensis* Rushton-Mellor, 1994
- *Argulus paranensis* Ringuelet, 1943
- *Argulus parsi* Tripathi, 1975
- *Argulus paulensis* C. B. Wilson, 1924
- *Argulus personatus* Cunnington, 1913
- *Argulus pestifer* Ringuelet, 1948
- *Argulus petagonicus* Ringuelet, 1943
- *Argulus phoxini* Leydig, 1871
- *Argulus piperatus* C. B. Wilson; 1920
- *Argulus pugettensis* Dana, 1852
- *Argulus puthenveliensis* Ramakrishna, 1959
- *Argulus quadristriatus* Devaraj & Ameer Hamsa, 1977
- *Argulus reticulatus* C. B. Wilson, 1920
- *Argulus rhamdiae* C. B. Wilson, 1936
- *Argulus rhipidiophorus* Monod, 1931
- *Argulus rijckmansii* Brian, 1940
- *Argulus rotundus* C. B. Wilson, 1944
- *Argulus rubescens* Cunnington, 1913
- *Argulus rubropunctatus* Cunnington, 1913
- *Argulus salminei* Kroyer, 1863
- *Argulus scutiformis* Thiele, 1900
- *Argulus shoutedeni* Monod, 1928
- *Argulus siamensis* C. B. Wilson, 1926
- *Argulus silvestrii* Lahille, 1926

- *Argulus smalei* Avenant-Oldewage & Oldewage, 1995
- *Argulus spinulosus* Silva, 1980
- *Argulus stizostethii* Kellicott, 1880
- *Argulus striatus* Cunnington, 1913
- *Argulus taliensis* Shen, 1948
- *Argulus tientsinensis* Ku & Wang, 1956
- *Argulus trachynoti* Brian, 1927
- *Argulus trilineatus* C. B. Wilson, 1904
- *Argulus varians* Bere, 1936
- *Argulus versicolor* C. B. Wilson, 1902
- *Argulus vierai* Pereira-Fonseca, 1939
- *Argulus violaceus* Thomsen, 1925
- *Argulus vittatus* (Rafinesque-Schmaltz, 1814)
- *Argulus wilsoni* Brian, 1940
- *Argulus yucatanus* Poly, 2005
- *Argulus yucatanus* Poly, 2004
- *Argulus yuii* K. N. Wang, 1958
- *Argulus yunnanensis* Shen, 1948

Binoculus Geoffroy St. Hilaire, 1762:

- *Binoculus bicornutus* Risso, 1816
- *Binoculus caudatus* Say, 1818
- *Binoculus gasterostei* Geoffroy, 1762
- *Binoculus haemisphericus* Geoffroy, 1762
- *Binoculus palustris* O. F. Müller, 1776
- *Binoculus piscinus* O. F. Müller, 1776
- *Binoculus productus* Nordmann, 1832
- *Binoculus salmoneus* O. F. Müller, 1785
- *Binoculus sexsetaceus* Nordmann, 1832

Chonopeltis Thiele, 1900:

- *Chonopeltis australis* Boxshall, 1976
- *Chonopeltis australissumus* Fryer, 1977
- *Chonopeltis brevis* Fryer, 1961
- *Chonopeltis congicus* Fryer, 1959
- *Chonopeltis elongatus* Fryer, 1974

- *Chonopeltis flaccifrons* Fryer, 1960
- *Chonopeltis fryeri* Van As, 1986
- *Chonopeltis inermis* Thiele, 1900
- *Chonopeltis koki* Van As, 1992
- *Chonopeltis lisikili* J. G. Van As & L. L. Van As, 1996
- *Chonopeltis meridionalis* Fryer, 1964
- *Chonopeltis minutus* Fryer, 1977
- *Chonopeltis schoutedeni* Brian, 1940
- *Chonopeltis victori* Avenant-Oldewage, 1991

Dipteropeltis Callman, 1912:

- *Dipteropeltis hirundo* Calman, 1912

Dolops Audouin, 1837:

- *Dolops bidentata* (Bouvier, 1899)
- *Dolops carvalhoi* Lemos de Castro, 1949
- *Dolops discoidalis* Bouvier, 1899
- *Dolops doradis* (Cornalia, 1860)
- *Dolops geayi* (Bouvier, 1897)
- *Dolops intermedia* da Silva, 1978
- *Dolops kollari* (Heller, 1857)
- *Dolops lacordairei* Audouin, 1837
- *Dolops longicauda* (Heller, 1857)
- *Dolops nana* Lemos de Castro, 1950
- *Dolops ranarum* (Stuhlmann, 1891)
- *Dolops reperta* (Bouvier, 1899)
- *Dolops striata* (Bouvier, 1899)
- *Dolops tasmanicus* Fryer, 1969

Huargulus Yü, 1938:

- *Huargulus chinensis* Yü, 1938

Impact

Fish lice occasionally reach high enough densities to cause fish kills in aquaculture operations, or more rarely in wild populations of fish. They can also become abundant in aquaria, sometimes resulting in the death of ornamental fish.

Ceratomyxa Shasta

Ceratomyxa shasta is a myxosporean parasite that infects salmonid fish on the Pacific coast of North America. It was first observed at the Crystal Lake Hatchery, Shasta County, California, and has now been reported from Idaho, Oregon, Washington, British Columbia and Alaska (Bartholomew et al. 1989).

Life History

In addition to the fish host, *C. shasta* infects a freshwater polychaete worm (Bartholomew et al., 1997). Actinospores are released from the worm, and infect fish, on contact, in the water column. Neither horizontal (fish to fish), nor vertical (fish to egg) transmissions have been documented under laboratory conditions, suggesting that the worm host is necessary for completion of the life cycle. Spores are released back into freshwater system after its fish host dies, however the complete life cycle, host and vector interaction is not fully understood (especially the ecology of the polychaete host). Research indicates that the potential for infection is enhanced when water temperatures are high, water flow is low, or numbers of infectious *C. shasta* are relatively high. Infection rates appear to be higher in or below still water environments than riverine ones.

Pathology of Infection

Clinical indications of infection in salmons include lethargy, loss of body mass, darkening of the skin, ascites, exopthalmia and kidney pustules, These symptoms vary from one salmonid species to another, and also depend on life stage of the host.

Internally, infection with *C. shasta* affects entire digestive tract, liver, gall bladder, spleen, gonads, kidney, heart, gills, and muscle tissues. Infection with *C. shasta* in adult chinook salmon causes mortality through intestinal perforations and co-occurring bacterial infections. Cold temperatures and salinity may reduce progress of disease, but do not eliminate infection. Progression of infection and mortality is temperature dependent, with higher temperature increasing disease progression and resulting in quicker mortality.

Disease Resistance

Salmonid stocks exhibit variable resistance to *C. shasta* (Bartholomew 1998). Resistance is variable and may be compromised by environmental conditions which enhance infectivity. Salmonid stocks which are resistant to *C. shasta* are not necessarily resistant to other myxosporean infections, such as *Myxobolus cerebralis*.

Dactylogyrus Vastator

Dactylogyrus vastator is a hermaphroditic Flatworm of class Monogenea which has been found to be an ectoparasite of fish. This flatworm is known to inhabit the gills of the fish and has cause problems with many fish farms. However, they remain mostly non-hazardous to humans.

Physical Characteristic

Worms of the species *D. vastator* are a little more than 1.25 millimeters long. They have two pairs of hooks known as hamuli. A set on the bottom surface of the worm is smaller than the set at the rear. These worms have three pairs of sticky sacs and four eyespots.

Distribution Range

D. vastator lives in the northern parts of North America, Europe, and Asia.

Significance to Humans

Causes mass mortality among fingerling carp in fish-rearing ponds. Abnormal multiplication of cells in the gill epithelium interferes with carp's respiratory function. The parasite is especially significant in the former Soviet Union and eastern, northern Europe, and India (Specifically Sri Lanka) where carp are bred for food.

Life Cycle

Dactylogyrus vastator has a life cycle with only 1 host, there are no intermediate hosts. New hosts are located and infected by free-living larvae (oncomiracidium). *D. vastator* lives in the gills of carp and goldfish. Adults on the gill filaments lay unembryonated eggs that are washed out of the host's gill cavity and sink to the bottom of the water. Ciliated oncomiracidium emerges in 3–5 days, depending on water temperature. Under summer temperatures, development of the embryo is rapid, and fully formed oncomiracidia appear in 2½ days. The larva presses from inside the egg and forces off one end of the egg shell, leaving the embryonic membrane inside when it escapes. Larvae are drawn into the gill cavity by the water current and attach themselves to another host's gills with their hooks. Larvae can also swim about actively and attach to the skin of the carp when it comes in contact with them. Having attached to the first host, they migrate over the body toward the gills, being attracted by the mucus from them. Sexual maturity is attained about 10 days after reaching the gills.

The peak of infestation on the fish is reached about the middle of July, after which the worms begin to disappear, being gone by fall. During the last phase of oviposition, winter eggs are laid; these remain in a resting stage (state of diapause, or period of inactivity) during the cold weather. With the arrival of spring and the warming of the water, development of the eggs begins and hatching soon takes place. Young carp especially are infested, often fatally when large numbers of worms are present on the gill filaments. Eggs laid during the summer hatch after a short period of incubation, resulting in an abundance of oncomiracidia with superinfection of the fish. Under these conditions, an immunity develops in the carp that reduces the infestation for about 2 months.

Diphyllobothrium

Diphyllobothrium is a genus of tapeworm which can cause Diphyllobothriasis in humans through consumption of raw or undercooked fish. The principal species causing diphyllobothriosis is *Diphyllobothrium latum*, known as the broad or fish tapeworm, or broad fish tapeworm. *D. latum* is a pseudophyllid cestode that infects fish and mammals. *D. latum* is native to Scandinavia, western Russia, and the Baltics, though it is now also present in North America, especially the Pacific Northwest. In Far East Russia, *D. klebanovskii*, having Pacific salmon as its second intermediate host, was identified. Other members of the genus *Diphyllobothrium* include *Diphyllobothrium dendriticum* (the salmon tapeworm), which has a much larger range (the whole northern hemisphere), *D. pacificum*, *D. cordatum*, *D. ursi*, *D. lanceolatum*, *D. dalliae*, and *D. yonagoensis*, all of which infect humans only infrequently. In Japan, the most common species in human infection is *D. nihonkaiense*, which was only identified as a separate species from *D. latum* in 1986. It was indicated to be synonymous to *D. klebanovskii* from the molecular study.

History

The fish tapeworm has a long documented history of infecting people who regularly consume fish and especially those whose customs include the consumption of raw or undercooked fish. In the 1970s, most of the known cases of diphyllobothriasis came from Europe (5 million cases), and Asia (4 million cases) with fewer cases coming from North America and South America, and no reliable data on cases from Africa or Australia. Interestingly, despite the relatively small number of cases seen today in South America, some of the earliest archeological evidence of diphyllobothriasis comes from sites in South America. Evidence of *Diphyllobothrium spp.* has been found in 4,000-10,000 year old human remains on the western coast of South America.

There is no clear point in time when *Diphyllobothrium latum* and related species were "discovered" in humans, but it is clear that diphyllobothriasis has been endemic in human populations for a very long time. Due to the changing dietary habits in many parts of the world, autochthonous, or locally-acquired, cases of diphyllobothriasis have recently been documented in previously non-endemic areas, such as Brazil. In this way, diphyllobothriasis represents an emerging infectious disease in certain parts of the world where cultural practices involving eating raw or undercooked fish are being introduced.

Morphology

The adult worm is composed of three fairly distinct morphological segments: the scolex (head), the neck, and the lower body. Each side of the scolex has a slit-like groove, which is a bothrium (tentacle) for attachment to the intestine. The scolex attaches to the neck, or proliferative region. From the neck, grows many proglottid segments which contain the reproductive organs of the worm. *D. latum* is the longest tapeworm in humans, averaging ten meters long. Adults can shed up to a million eggs a day.

In adults, proglottids are wider than they are long (hence the name *broad tapeworm*). As in all pseudophyllid cestodes, the genital pores open midventrally.

Life Cycle

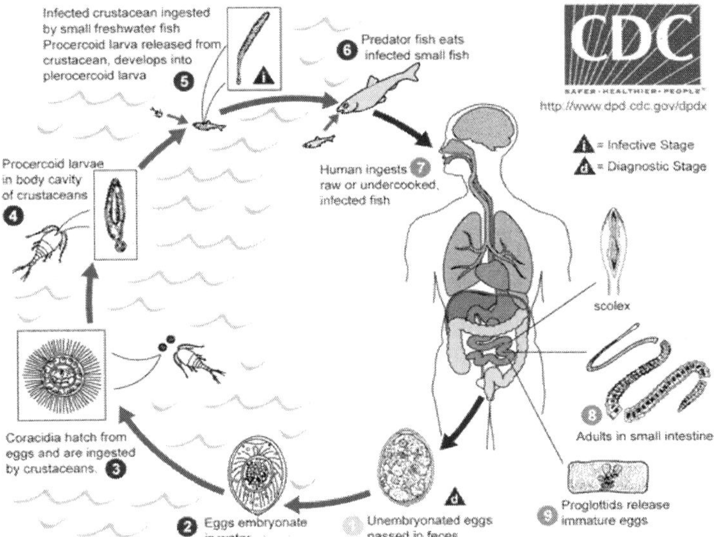

***Figure:** Life cycle of* D. latum. *Click the image to see full-size.*

Adult tapeworms may infect humans, canids, felines, bears, pinnipeds, and mustelids, though the accuracy of the records for some of the nonhuman species is disputed. Immature eggs are passed in feces of the mammal host (the definitive host, where the worms reproduce). After ingestion by a suitable freshwater crustacean such as a copepod (the first intermediate host), the coracidia develop into procercoid larvae. Following ingestion of the copepod by a suitable second intermediate host, typically a minnow or other small freshwater fish, the procercoid larvae are released from the crustacean and migrate into the fish's flesh where they develop into a plerocercoid larvae (sparganum). The plerocercoid larvae are the infective stage for the definitive host (including humans).

Because humans do not generally eat undercooked minnows and similar small freshwater fish, these do not represent an important source of infection. Nevertheless, these small second intermediate hosts can be eaten by larger predator species, for example, trout, perch, and walleyed pike. In this case, the sparganum can migrate to the musculature of the larger predator fish and mammals can acquire the disease by eating these later intermediate infected host fish raw or undercooked. After ingestion of the infected fish, the plerocercoids develop into immature adults and then into mature adult tapeworms which will reside in the small intestine. The adults attach to the intestinal mucosa by means of the two bilateral grooves (bothria) of their scolex. The adults can reach more than 10 m (up to 30 ft) in length in some species such as *D. latum,* with more than 3,000 proglottids. One or several of the tape-like proglottid segments (hence the name tape-worm) regularly detach from the main body of the worm and release immature eggs in fresh water to start the cycle over again. Immature eggs are discharged from the proglottids (up to 1,000,000 eggs per day per worm) and are passed in the feces. The incubation period in humans, after which eggs begin to appear in the feces is typically 4–6 weeks, but can vary from as short as 2 weeks to as long as 2 years. The tapeworm can live up to 20 years.

Clinical Symptoms, Including Occasional Parasite-induced B_{12} Deficiency

Symptoms of diphyllobothriasis are generally mild, and can include diarrhea, abdominal pain, vomiting, weight loss, fatigue, constipation and discomfort. Approximately four out of five cases are asymptomatic and may go many years without being detected. In a small number of cases, this leads to severe vitamin B_{12} deficiency due to the parasite absorbing 80% or more of the host's B_{12} intake, and a megaloblastic anemia indistinguishable from pernicious anemia. The anemia can also lead to subtle demyelinative neurological symptoms (subacute combined degeneration of spinal cord). Infection for many years is ordinarily required to deplete the human body of vitamin B-12 to the point that neurological symptoms appear.

Diagnosis

Diagnosis is usually made by identifying proglottid segments, or characteristic eggs in the feces. These simple diagnostic techniques are able to identify the nature of the infection to the genus level, which is usually sufficient in a clinical setting. However, when the species needs to be determined (in epidemiological studies, for example), restriction fragment length polymorphisms can be effectively used. PCR can be performed on samples of purified eggs, or native fecal samples following sonication of the eggs to release their contents.

Treatment

Upon diagnosis, treatment is quite simple and effective. The standard treatment for diphyllobothriasis, as well as many other tapeworm infections is a single dose of Praziquantel, 5–10 mg/kg PO once for both adults and children. An alternative treatment is Niclosamide, 2 g PO once for adults or 50 mg/kg PO once. Another interesting potential diagnostic tool and treatment is the contrast medium, Gastrografin, introduced into the duodenum, which allows both visualization of the parasite, and has also been shown to cause detachment and passing of the whole worm.

Side Effects of Treatment

Praziquantel has few side effects, many of which are similar to the symptoms of diphyllobothriasis. They include malaise, headache, dizziness, abdominal discomfort, nausea, rise in temperature and occasionally allergic skin reactions. The side effects of Niclosamide are very rare, due to the fact that it is not absorbed in the gastrointestinal tract.

Epidemiology

People at high risk for infection have traditionally been those who regularly consume raw fish, including fishermen who eat the raw liver or roe of their catches and women preparing and tasting foods that contain raw fish. Many regional cuisines include raw or undercooked food, including sushi and sashimi in Japanese cuisine, carpaccio di persico in Italian, tartare maison in French-speaking populations, ceviche in Latin American cuisine and marinated herring in Scandinavia. With emigration and globalization, the practice of eating raw fish in these and other dishes has brought diphyllobothriasis to new parts of the world and created new endemic foci of disease.

Public Health Strategies

The most viable interventions include: prevention of water contamination both by raising public awareness of the dangers of defecating in recreational bodies of water and by implementation of basic sanitation measures; screening and successful treatment of people infected with the parasite; and prevention of infection of humans via consumption of raw, infected fish. The last of these can most easily be changed via education about proper preparation of fish. Fish that is thoroughly cooked, brined, or frozen at -10°C for 24–48 hours can be consumed without risk of D. latum infection.

Epizootic Ulcerative Syndrome

Epizootic ulcerative syndrome (EUS) is a disease caused by the fungus, *Aphanomyces invadans*. It infects many freshwater and brackish fish species in the Asia-Pacific region and Australia. The disease is most commonly seen when there are low temperature and heavy rainfall in tropical and sub-tropical waters. It may also be referred to as Mycotic Granulomatoses (MG) and Red Spot Disease (RSD).

Clinical Signs and Diagnosis

At first, fish develop red spots on the skin. These lesions expand to form ulcers and extensive erosions filled with necrotic tissue and mycelium. This is followed by the development of granulomas on the internal organs and death. A provisional diagnosis can be made by using a squash preparations of the skeletal muscle from beneath an ulcer to identify septate fungal hyphae. Definitive diagnosis can be made based on histopathogical findings and fungal isolation.

Treatment and Control

Infected fish should be moved into high quality water, where they may recover if their clinical signs are mild. If disease occurs eradication is required. Once the disease is eradicated good husbandry, surveillance and biosecurity measures are necessary to prevent recurrence. In countries free of Epizootic Ulcerative Syndrome, quarantine and health certificates are necessary for the movement of all live fish to prevent the introduction of the disease.

Trematoda

Trematoda is a class within the phylum Platyhelminthes that contains two groups of parasitic flatworms, commonly referred to as "flukes".

Taxonomy and Biodiversity

The trematodes or flukes are estimated to include 18,000 to 24,000 species, and are divided into two subclasses. Nearly all trematodes are parasites of mollusks and vertebrates. The smaller Aspidogastrea, comprising about 100 species, are obligate parasites of mollusks and may also infect turtles and fish, including cartilaginous fish. The Digenea, which constitute the majority of trematode diversity, are obligate parasites of both mollusks and vertebrates, but rarely occur in cartilaginous fish. Formerly the Monogenea were included in Trematoda on the basis that these worms are also vermiform parasites, but modern phylogenetic studies have raised this group to the status of a sister class within the Platyhelminthes, along with the Cestoda.

Anatomy

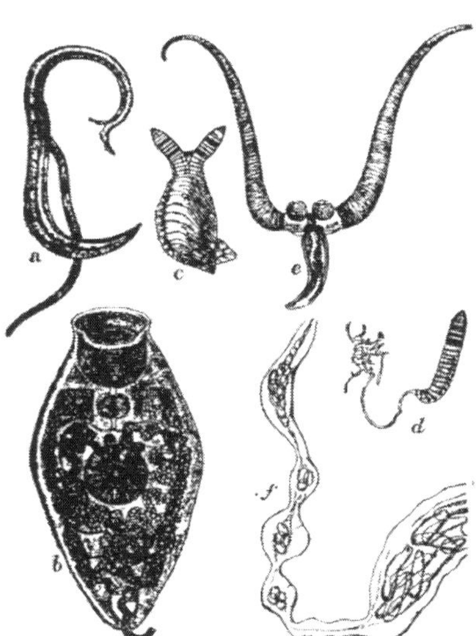

Figure: *Varied trematodes, from 1911 Encyclopaedia Britannica*

Trematodes are flattened oval or worm-like animals, usually no more than a few centimetres in length, although species as small as 1 millimetre (0.039 in) and as large as 7 metres (23 ft) are known. Their most distinctive external feature is the presence of two suckers, one close to the mouth, and the other on the underside of the animal.

The body surface of trematodes comprises a tough syncitial tegument, which helps protect against digestive enzymes in those species that inhabit the gut of larger animals. It is also the surface of gas exchange; there are no respiratory organs.

The mouth is located at the forward end of the animal, and opens into a muscular, pumping pharynx. The pharynx connects, via a short oesophagus, to one or two blind-ending caeca, which occupy most of the length of the body. In some species, the caeca are themselves branched. As in other flatworms, there is no anus, and waste material must be egested through the mouth.

Although the excretion of nitrogenous waste occurs mostly through the tegument, trematodes do possess an excretory system, which is instead mainly concerned with osmoregulation. This consists of two or more protonephridia, with those on each side of the body opening into a collecting duct. The two collecting ducts typically meet up at a single bladder, opening to the exterior through one or two pores near the posterior end of the animal.

The brain consists of a pair of ganglia in the head region, from which two or three pairs of nerve cords run down the length of the body. The nerve cords running along the ventral surface are always the largest, while the dorsal cords are present only in the Aspidogastrea. Trematodes generally lack any specialised sense organs, although some ectoparasitic species do possess one or two pairs of simple ocelli.

Reproductive System

Most trematodes are simultaneous hermaphrodites, having both male and female organs. There are usually two testes, with sperm ducts that join together on the underside of the front half of the animal. This final part of the male system varies considerably in structure between species, but may include sperm storage sacs and accessory glands, in addition to the copulatory organ, which is either eversible, and termed a *cirrus*, or non-eversible, and termed a penis.

There is usually only a single ovary, which is connected, via a pair of ducts to a number of *vitelline glands* on either side of the body, that produce yolk cells. Eggs pass from the ovary into a glandular receptacle called the *ootype* or *Mehlis' gland*, where fertilization occurs. This opens into an elongated uterus that opens to the exterior close to the male opening. The ovary is often also associated with a storage sac for sperm, and a copulatory duct termed *Laurer's canal.*

Life Cycles

Almost all trematodes infect mollusks as the first host in the life cycle, and most have a complex life cycle involving other hosts. Most trematodes are monoecious and alternately reproduce sexually and asexually. The two main exceptions to this are the Aspidogastrea, which have no asexual reproduction, and the schistosomes, which are dioecious.

In the definitive host, in which sexual reproduction occurs, eggs are commonly shed along with host feces. Eggs shed in water release free-swimming larval forms that are infective to the intermediate host, in which asexual reproduction occurs.

A species that exemplifies the remarkable life history of the trematodes is the bird fluke, *Leucochloridium paradoxum*. The definitive hosts, in which the parasite multiplies, are various woodland birds, while the hosts in which the parasite grows (intermediate host) are various species of snail. The adult parasite in the bird's gut produces eggs and these eventually end up on the ground in the bird's faeces. Some very fortunate eggs get swallowed by a snail and here they hatch into tiny, transparent larva (miracidium). These larvae grow and take on a sac-like appearance. This stage is known as the sporocyst and it forms a central body in the snail's digestive gland that extends into a brood sac in the snail's head, muscular foot and eye-stalks. It is in the central body of the sporocyst where the parasite replicates itself, producing lots of tiny embryos (redia). These embryos move to the brood sac and mature into cercaria.

Infections

Human infections are most common in Asia, Africa, South America, or the Middle East. However, trematodes can be found anywhere that human waste is used as fertilizer.

Etymology

Trematodes are commonly referred to as *flukes*. This term can be traced back to the Old English name for flounder, and refers to the flattened, rhomboidal shape of the worms.

The flukes can be classified into two groups, on the basis of the system which they infect in the vertebrate host.

- Tissue flukes infect the bile ducts, lungs, or other biological tissues. This group includes the lung fluke, *Paragonimus westermani*, and the liver flukes, *Clonorchis sinensis* and *Fasciola hepatica*.

- Blood flukes inhabit the blood in some stages of their life cycle. *Blood flukes include species of the genus* Schistosoma.

They may also be classified according to the environment in which they are found. For instance, pond flukes infect fish in ponds.

Glugea

Glugea is a genus of microsporidian parasites, predominantly infecting fish.

Species include

- *Glugea anomala*
- *Glugea atherinae*
- *Glugea capverdensis* - a parasite of the fish *Myctophum punctatum*
- *Glugea caulleryi* - a parasite of the greater sand eel *Hyperoplus lanceolatus* (a teleost fish)
- *Glugea heraldi* - a parasite of the seahorse *Hippocampus erectus*
- *Glugea hertwigi* - a parasite of the smelts *Osmerus eperlanus* and *Osmerus mordax*
- *Glugea merluccii* - a parasite of the fish *Merluccius hubbsi*
- *Glugea plecoglossi*
- *Glugea shiplei* - a parasite of fish of the genus *Trisopterus*
- *Glugea stephani* - a parasite of the winter flounder *Pseudopleuronectes americanus* (a pleuronectid flatfish)
- *Glugea vincentiae* - a parasite of the Australian marine teleost fish, *Vincentia conspersa*
- *Glugea weissenbergi*

Glugea americanus may be better placed in *Spraguea*

Gyrodactylus Salaries

Gyrodactylus salaris is a small monogenean ectoparasite (about 0.5 mm long) which mainly lives on the skin of freshwater fish, especially Atlantic salmon. This species is a leech-like parasite that has been implicated in some diminution of Atlantic salmon populations in the Norwegian fjords. Its common name is Salmon Fluke. Other species that can be parasitized include rainbow trout, Arctic char, North American brook trout, grayling, North American lake trout and brown trout.

The parasite attaches to the fish by a large specialized posterior attachment organ, the (haptor) which has sixteen sharp hooks located around its margin. The parasite cannot be seen with the naked eye, but it can be seen with a hand held lens. When feeding, the parasite attaches its anterior end to the fish with cephalic glands. It everts its pharynx through the mouth and releases a digestive solution with proteolytic enzymes which dissolves the salmon skin. Mucus and dissolved skin are then sucked into the gut. Attachment of many parasites can cause large wounds and the epidermis of the host fish can be damaged which allows in secondary infections. The parasites gives birth to live young nearly as big as themselves and at this time, a further generation is already growing inside the neonates.

Catastrophic losses of Atlantic salmon occurred in Norway in the 1970s following the introduction of *G. salaris*. By 2001, the salmon populations of 41 Norwegian rivers had been virtually wiped out in this way.

The parasite cannot survive in the high salinity of sea water, so the infection is not spread by the migration of fish. Historically, *Gyrodactylus*-infected rivers have been treated with the indiscriminate pesticide/piscicide rotenone. A newer method of treatment employs dosing small volumes of aqueous aluminium and sulfuric acid into the river. A huge advantage of this method is its ability to kill the parasites without harming the hosts. This new method has shown promising results in Batnfjordselva River and Lærdalselva River, two rivers in Norway.

Henneguya Zschokkei

Henneguya zschokkei is species of a myxosporean parasite of certain species of salmons of genus *Oncorhynchus*.

Synonyms:

- *Henneguya salminicola* Ward, 1919

Hosts:

- *Oncorhynchus nerka*
- *Oncorhynchus keta*
- *Oncorhynchus tshawytscha*
- *Oncorhynchus gorbuscha*
- *Oncorhynchus kisutch*
- migrating form of *Oncorhynchus mykiss*

8

Lymphocystis Disease of Fishes

Description

Lymphocystis is an infectious viral disease of freshwater and saltwater fishes that causes cell enlargement (many more times than normal cell size), also called hypertrophy, usually on the skin and fins. The enlarged cell nodules may each reach 0.3 mm to greater than 2.0 mm in diameter, each becoming a virus factory. Lymphocystis is the most common viral infection of aquarium fish, and has been reported in over 125 species of freshwater and marine fishes.

The disease usually runs its course in 4 or more weeks (depending on species involved, water temperature, and other variables) and then the enlarged cells rupture or slough off and release the viral particles into the water. While infected, the fish may become slowed or weakened, or more visible, and thus be more prone to predation or attack. If there are mouth lesions, the fish may have difficulty in feeding or may not be able to feed. The low mortality rate some attribute to lymphocystis is mostly due to secondary bacterial or fungal infections. I have worked with many thousands of fishes of various freshwater and salt water species and do not recall a single death being directly due to a lymphocystis infection.

After lymphocystis lesions are lost, the host tissue heals up. Adhesions and scarring can occur during healing. If the gills are affected, the fish can have difficulty breathing, especially if gill surface areas are destroyed (no longer present), or adhesions or scarring occur and gill surfaces are thus reduced in surface area or functional quality for oxygen uptake.

The viral particles in the water can go on to infect another fish of the same or closely related species. I suspect the virus can also become dormant and remain viable in sediments for years. The virus may be stored for years (for future research) by either freezing or freeze-drying the separated viral particles, infected tissues, or infected whole fish. The disease poses no known health hazard to humans. To decrease spreading this disease to other fishes, infected fish should be buried or burned, and not thrown back into the water. The viral agent of lymphocystis disease is an iridovirus (of the family Iridoviridae) called *Lymphocystivirus* (genus name). Iridoviruses range from 120-300 nm (nanometer = one billionth of a meter) in diameter. Iridoviruses have an icosohedral or 20-sided shape, and a DNA core.

I have worked on the disease in silver perch (*Bairdiella chrysura*), white-tailed damselfish (*Dascyllus aruanus*), black-tailed humbug (*Dascyllus melanurus*), copper banded angelfish (*Chelmon rostratus*), Koran angelfish (*Pomacanthus semicirculatus*), Moorish idol (*Zanclus canescens*), foureye butterflyfish (*Chaetodon capistratus*), orbiculate bat fish (*Platax orbicularis*), queen angelfish (*Holacanthus ciliaris*), and warmouth or goggle-eye (*Lepomis gulosus*).

DNA studies have showed that there are different species of the virus. This has been suspected for some time because the viral particles from different fishes vary in size plus the virus from a fish usually will infect only that species of fish or a few other species closely related to the primary host.

The virus enters through broken skin or injured tissue (usually skin or fin). If the virus gets into the blood (usually via gill infections) then various internal organs can be infected. In 1974 I showed that the spleen, tissues behind the eye, eye, and many other internal organs can be infected via systemic infection. One can easily infect fish by putting them into a bucket of water, introducing the virus, then injuring fish by vigorously swirling a stiff bottle brush in the bucket. One can also run a sharp probe on the skin or tail and expose the fish to the virus to infect them.

Incubation times (until lesions are visible to eye) range from about 10 days at 25 C to longer, depending on species involved, temperature, and other variables.

Important or valuable affected fish should be isolated and monitored for secondary bacterial or fungal infections that should be treated with appropriate drugs.

In 1979 I discovered that goggle-eye (warmouth), when subjected to heavy rains and sediment loads, came down with lymphocystis. It is unknown, in this case, whether stress or physical injury led to the lymphocystis infections. Stress to the fish, in this case, could be from exposure to sediments in the water leading to breathing problems, from getting tired trying to maintain their position in the swift water, from being exposed to toxins swept in by the water, etc. There is also the possibility that sediment or debris particles hitting the fish in the swiftly moving water caused injury that led to the infections (similar to injuring fish in a bucket by swirling a brush).

Control

There is no way to cure a viral disease in any organism yet, no matter what some of the fish medications claim. Some virus diseases in various organisms can be prevented by vaccination, or slowed down by medications. There is no vaccine for lymphocystis yet.

Some medications claim to cure lymphocystis in several days to a week. However, depending on the stage of the disease, many enlarged cells will burst and disappear on their own (without using the "cure"), making some think their chemical has cured the disease. One can also excise the enlarged cells, affecting an apparent "cure," but, in reality, just removing most of the infected cells (which then can no longer be seen). Lymphocystis cannot be "cured."

Lymphocystis, once acquired, must run its course. There are some ways to decrease the chance of fish getting the disease:

...Handle fish carefully to reduce injury to skin and fins and slime coat.

...Reduce or prevent stress on the fish. Stress would include overcrowding, starvation, overfeeding, sudden environmental changes (temperature, salinity, pH, DO, strong currents carrying debris, etc), toxins (pesticides used nearby; high ammonia, nitrite, or nitrate), or anything else that could affect the immune system or cause physical injury. Lymphocystis frequently appears on the tail due to nipping by other fish, usually a result from overcrowding.

While I was working on this disease in silver perch, I suddenly had numerous warts appear on my hands. I handled many experimentally infected fish. I could get no doctor to excise a wart and compare the wart virus with the lymphocystis virus I was working on, so I leave the answer to this question to those in the future. Several months later, shortly after I worked in concrete (a basic pH medium) building a concrete pond, the warts just as mysteriously disappeared, and remain gone over 30 years later.

Did I find that lymphocystis can infect humans? And that warts can be cured by abrasion and/or a basic solution? Only the future work of others will tell.

European Community Reference Laboratory for Fish Diseases

The European Community Reference Laboratory for Fish Diseases is currently located in Aarhus in Denmark at the National Veterinary Institute (a part of Technical University of Denmark).

A main purpose of the CRL is to ensure the quality of diagnostics of fish diseases in EU Member States and to harmonise the procedures and methodologies applied. The work is mainly concerned with the exotic (Epizootic Haematopoietic Necrosis Virus and Epizootic Ulcerative Syndrome) and non-exotic (Viral hemorrhagic septicemia, Infectious hematopoietic necrosis virus, Koi herpes virus, and Infectious salmon anemia virus) diseases mentioned in Council Directive 2006/88/EC. The CRL co-ordinates those activities of the National Reference Laboratories (NRLs) for Fish diseases in EU that aim to harmonise diagnostic techniques and disseminate information of mutual interest. The work of the CRL also includes development of a database (Fishpathogens.eu) that aims to collect information on fish disease isolates and their sequences. Details of the work programme is decided at the Annual Meeting of the NRLs for Fish Diseases.

Lymphocystivirus

Lymphocystivirus is one of five genera of viruses within the viral family *Iridoviridae*, and one of three genera within this family which infect teleost fishes, along with *Megalocytivirus* and *Ranavirus*. Lymphocystiviruses infect more than 140 freshwater and marine species, spanning at least 42 host families worldwide, causing the chronic, self-limiting clinical disease, lymphocystis. While lymphocystis does not cause mass mortality events like megalocytiviruses and ranaviruses, fish with lymphocystis exhibit grossly visible papilloma-like skin lesions which substantially reduce their commercial value. No vaccines are currently available for lymphocystis viruses.

Taxonomy

The genus *Lymphocystivirus* has at least two viral species: Lymphocystis disease virus 1 (LCDV-1) and Lymphocystis disease virus 2 (LCD-2). LCDV-1 infects European flounder *Platichthys flesus* and European plaice (*Pleuronectes platessa*). LCDV-2 infects the common dab (*Limanda limanda*). A third species Lymphocystis disease virus C has also been proposed based on the relative lack of sequence similarity of an isolate from the olive flounder (*Paralichthys olivaceous*) to LCDV-1 and LCDV-2.

LCDV Genome

Lymphocystiviruses are Group I viruses with a dsDNA genome. The LCDV-1 genome is approximately 102.7 kilobase pairs (kbp) in length, with 195 potential open reading frames (ORF), and codes for two DNA-dependent RNA polymerase subunits, a DNA methyltransferase, a DNA polymerase, a guanosine triphosphate phosphohydrolase (GTPase), a helicase, protein kinases, a ribonucleoside diphosphate reductase, and zinc-finger proteins, among others. The LCDV-2 genome is similar to that of LCDV-1 but is slightly smaller, approximately 98 kilobase pairs (kbp) in length.

Structure and Replication

Megalocytiviruses are large icosahedral DNA viruses measuring from 198-227 nm in diameter (in some cases as large as 380) nm in diameter with a large single linear dsDNA genome. The main structural component of the protein capsid is the major capsid protein (MCP)

Lymphocystiviruses attach to the host cell and enter by receptor-mediated endocytosis similar to other iridoviruses. Viral particles are uncoated and move to the nucleus of the cell, where DNA replication begins via a virally encoded DNA polymerase. Viral DNA then moves to the cytoplasm for the second stage of DNA replication, which results in the formation of DNA concatemers. The concatameric viral DNA is subsequently packaged via a headful mechanism into virions. The lymphocystis viral genome is circularly permuted with terminally redundant DNA.

Pathogenesis

Lymphocystis disease is a chronic disease that rarely causes mortality. Infection causes transformation and hypertrophy (approximately 1000x) of cells in the dermis, forming grossly visible lymphocystis nodules, as well as transformation and hypertrophy in cells of the connective tissues of various internal organs. Fibroblasts and osteoblasts are specifically targeted by the virus. Lymphocystis viruses are not easily grown in cell culture, placing limitations on *in vitro* molecular pathogenesis experiments.

Diagnostic Pathology

As Lymphocystis viruses are not easily grown in cell culture, diagnosis is based on clinical signs, gross pathology, histopathology, serology, and/ or polymerase chain reaction (PCR)-based molecular assays.

Gross Pathology

The pathology of lymphocystis consists of papilloma-like skin lesions composed of greatly hypertrophied infected host cells embedded in extracellular matrix, sometimes called lymphocystis tumor cells, which are grossly evident as white spots on the skin and fins of infected fish. These lesions proliferate as epithelial tumors in some cases.

Histopathology

In a recent comparison of lymphocystis histopathology of four unrelated marine species, lesions consistently associated with lymphocystis included hypertrophied cells displaying irregular nuclei, basophilic cytoplasmic inclusion bodies that stained positively via Feulgen and Mann's reaction and Periodic acid-Shiff (PAS)-positive hyaline capsules. Hyaline capsules arise from the extracellular matrix that is produced by the infected cells, and are composed of sulphated and carboxylated glycoproteins (acid mucopolysaccharides). In contrast, the inclusion body shape, distribution of viral particles within the cytoplasm and overall appearance of lymphocystis nodules varied by species. The species examined in this study included the the white-spotted puffer (*Arothron hispidus*), the Japanese sea bass (*Lateolabrax japonicus*), olive flounder (*Paralichthys olivaceus*) and the "sting fish" or Schlegel's black rockfish (*Sebastes schegeli*)

Serology

Several serologic assays have been developed to identify LCDV infections, including flow cytometry, immunoblot, and immunofluorescence. However, PCR-based molecular assays are more practical for most applications.

Electron Microscopy

Transmission electron microscopy (TEM) of infected cells reveals cytoplasmic virus particles typically measuring from 198-227 nm in diameter (in some cases as large as 380 nm) and electron-dense substances in the perinuclear space.

Molecular Pathology

Published PCR primers and protocol are available to amplify a portion of the LCDV-1 MCP. When the PCR diagnostic assay is combined with slot blot, diagnostic sensitivity is increased, facilitating the diagnosis of asymptomatic LCDV-1 infections.

M74 Syndrome

M74 syndrome is a thiamine responsive disease of salmon (*Salmo salar*) in the Baltic Sea and neighboring waters which leads to the death of nearly all fry of certain females.

Megalocytivirus

Megalocytivirus is one of five genera of viruses within the family Iridoviridae and one of three genera within this family which infect teleost fishes, along with Lymphocystivirus and Ranavirus. The megalocytiviruses are an emerging group of closely related dsDNA viruses which cause systemic infections in a wide variety of wild and cultured fresh and saltwater fishes. Megalocytivirus outbreaks are of considerable economic importance in aquaculture, as epizootics can result in moderate fish loss or mass mortality events of cultured fishes.

Taxonomy

The family Iridoviridae is divided into five genera which include Chloriridovirus, Iridovirus, Lymphocystivirus, Megalocytivirus, and Ranavirus. Megalocytivirus is the most recently added genus. Megalocytivirus isolates exhibit relatively few genetic differences and have been divided into three major groups based on genetic sequence data; these groups are represented by infectious spleen and kidney necrosis virus (ISKNV), red sea bream iridovirus (RSIV), and turbot reddish body iridovirus (TRBIV). RSIV and ISKNV are the best known of the megalocytiviruses.

Song, *et al.* evaluated 48 Asian and Australian megalocytivirus isolates with regard to geographic location and genetic variation in the major capsid protein gene and developed a phylogenetic tree which divided the 48 isolates into three distinct clusters based on genotype. One of these clusters (genotype I) is widely distributed among several Asian countries, including 13 isolates from Korea, nine isolates from Japan, one from Thailand, one from China, and one from the South China Sea. In contrast, the other two genotypes had a smaller host range and were locally distributed. Genotype II megalocytiviruses infected freshwater fishes from Southeast Asia and Australia, whereas genotype III megalocytiviruses infected primarily flatfish in China and Korea.

Structure and Replication

Megalocytiviruses are large icosahedral DNA viruses measuring 150-250 nm in diameter with a large single linear dsDNA genome.

Megaloviruses are assumed to replicate in the same fashion as other Iridoviruses, and attach to the host cell and enter by receptor-mediated endocytosis.

Uncoated viral particles subsequently translocate to the host cell nucleus, where a virally encoded DNA polymerase facilitates DNA replication. Viral DNA then leaves the nucleus of the host cell and a second stage of DNA replication occurs in the cytoplasm, forming DNA concatemers.

A headful mechanism is utilized to package the concatameric viral DNA into virions formed at cytoplasmic virus assembly sites. Iridoviral DNA, unlike other DNA viruses infecting eukaryotic cells, is circularly permuted and exhibits terminal redundancies.

Transmission and epizoology

Transmission of megalocytivirus is believed to occur when a naive fish ingests tissues from infected fish or via contaminated water. Considerable effort has been expended to understand the transmission and epizoology of megalocytiviruses because of the economic importance of commercial fisheries and aquaculture operations. Interestingly, iridoviral epizootics do not correlate well with commercial food fish trade routes, with the notable exception of larval fish trade in Korea and Japan.

A second potential mechanism for accidental movement of infected fish is the international trade in ornamental or aquarium fishes, which includes the global trade of approximately 5000 freshwater and 1450 saltwater fishes. Each year over 1 billion individual fish are shipped among more than 100 nations, creating a serious concern for the spread of megalocytiviruses as well as other important fish pathogens. There is already substantial evidence of this problem: megalocytiviruses which are genetically identical or exremely similar to ISKNV have been isolated from ornamental fishes (gouramis) that were being traded internationally. Furthermore, an Australian outbreak of megalocytivirus among farmed Murray cod (*Maccullochella peelii*) was linked to imported gouramis in pet shops. In addition, a 2008 study reported 10 aquarium fish species that tested positive for ISKNV in Korea.

Pathogenesis

As megalocytiviruses have only been recently identified and described, the pathogenesis of megalocytivirus infection is relatively poorly understood. Clinical signs associated with infection are nonspecific and may include appetite loss, uncoordinated swimming, lethargy, coelomic distention, darkening skin color, petechiae, fin erosion, and death.

Large conspicuous hypertrophied cells, for which the genus is named, are evident in multiple organs when diseased tissues are examined by histopathology; these distinctive cells are commonly observed in the kidney, spleen and gastrointestinal tract and less commonly seen in the liver, gills, heart, and connective tissue. The hypertrophied cells are frequently perivascular in distribution and are greatly enlarged due to large granular to foamy basophilic cytoplasmic inclusion bodies. If the distended cells occlude the vasculature, focal areas of ischemic necrosis may be evident within various organs. It has been suggested by some researchers that the hypertrophied cells are some type of leukocyte, which is consistent with their tissue distribution.

Mesnilium

Mesnilium is a genus of parasitic protozoa of belonging to the phylum Apicomplexia. Its vertebrate hosts are fish. The vectors are not presently known but are thought likely to be leeches.

History

The genus was created in 1972 by Misra, Haldar and Chakravarty.

Description

Merogony occurs in erythrocytes and reticulo-endothelial cells.

Gamogony occurs only in erythrocytes.

Pigment granules are present only in microgamonts and macrogametes.

Hosts

The only known host of this parasite is the freshwater murrel (*Ophicephalus punctatus*).

Monogenea

Monogenea (adj. monogenean) are a group of largely ectoparasitic members of the flatworm phylum Platyhelminthes, class Monogenea.

Characteristics

Monogenea are very small parasitic flatworms mainly found on skin or gills of fish. They are rarely longer than about 2 cm. A few species infecting certain marine fish are larger and marine forms are generally larger than those found on fresh water hosts. Monogeneans lack respiratory, skeletal and circulatory systems and have no or weakly developed oral suckers. Monogenea attach to hosts using hooks, clamps and a variety of other specialized structures. They are often capable of dramatically elongating and shortening as they move. Biologists need to ensure that specimens are completely relaxed before measurements are taken.

Like all ectoparasites, monogeneans have well-developed attachment structures. The anterior structures are collectively termed the prohaptor, while the posterior ones are collectively termed the opisthaptor. The posterior opishaptor with its hooks, anchors, clamps etc. is typically the major attachment organ.

Like other flatworms, Monogenea have no true body cavity (coelom). They have a simple digestive system consisting of a mouth opening with a muscular pharynx and an intestine with no terminal opening (anus). Generally, they also are hermaphroditic with functional reproductive organs of both sexes occurring in one individual. Most species are oviparous but a few are viviparous. Monogenea are Platyhelminthes and therefore are among the lowest invertebrates to possess three embryonic germ layers—endoderm, mesoderm, and ectoderm. In addition, they have a head region that contains concentrated sense organs and nervous tissue (brain).

Systematics and Evolution

The ancestors of Monogenea were probably free-living flatworms similar to modern Turbellaria. According to the more widely accepted view, "rhabdocoel turbellarians gave rise to monogeneans; these, in turn, gave rise to digeneans, from which the cestodes were derived. Another view is that the rhabdocoel ancestor gave rise to two lines; one gave rise to monogeneans, who gave rise to digeneans, and the other line gave rise to cestodes" .

There are about 50 families and thousands of described species.

Some parasitologists divide Monogenea into two (or three) subclasses based on the complexity of their haptor: Monopisthocotylea have one main part to the haptor, often with hooks or a large attachment disc, whereas Polyopisthocotylea have multiple parts to the haptor, typically clamps. These groups are also known as Polyonchoinea and Heteronchoinea, respectively. Polyopisthocotyleans are almost exclusively gill-dwelling blood feeders, whereas Monopisthocotyleans may live on the gills, skin and fins.

Monopistocotylea include:

- Genus *Gyrodactylus,* which has no eyespots and is viviparous.
- Genus *Dactylogyrus,* which has four eyespots and is oviparous. This is one of the largest metazoan genera, with at least 970 species.
- Genus *Neobenedenia,* which is much larger and lives on the skin of many tropical marine species, causing problematic infections in marine aquaria.

All of which can cause epizootics in freshwater fish when raised in aquaculture.

Polyopisthocotylea include:

- Genus *Diclidophora,* which is primarily found in marine fish and primitive freshwater fish like sturgeons and paddlefish.
- Genus *Protopolystoma,* found in aquatic clawed toads (*Xenopus* species).

Ecology and Life Cycle

Monogeneans possess the simplest life cycle among the parasitic platyhelminths. They have no intermediate hosts and are ectoparasitic on fish (seldom in the urinary bladder and rectum of cold-blooded vertebrates). Although they are hermaphrodites, the male reproductive system becomes functional before the female part. The eggs hatch releasing a heavily ciliated larval stage known as an *oncomiracidium*. The *oncomiracidium* has numerous posterior hooks and is generally the life stage responsible for transmission from host to host. No known monogeneans infect birds, but one (*Oculotrema hippopotami*) infects mammals, parasitizing the eye of the hippopotamus.

Mononegavirales

The order *Mononegavirales* is the taxonomic home of numerous related viruses. Members of the order that are commonly known are, for instance, Ebola virus, human respiratory syncytial virus, measles virus, mumps virus, Nipah virus, and rabies virus. All of these viruses cause significant disease in humans. Many very important pathogens of nonhuman animals and plants are also members of this order.

Use of Term

The order *Mononegavirales* is a virological taxon (i.e. a concept) that was created in 1991 and emended in 1995, 1997, 2000, and 2005. The name *Mononegavirales* is derived from the Greek adjective *monos* (alluding to the single-stranded genomes of mononegaviruses), the Latin verb *negare* (alluding to the negative polarity of these genomes], and the taxonomic suffix *-virales* (denoting a viral order). The order currently includes the four virus families *Bornaviridae, Filoviridae, Paramyxoviridae*, and *Rhabdoviridae*.

Note

Mononegavirales is pronounced IPA or mo-nuh-ne-guh-vee-rah-liz in English phonetic notation. According to the rules for taxon naming established by the International Committee on Taxonomy of Viruses (ICTV), the name *Mononegavirales* is always to be capitalized, italicized, never abbreviated and to be preceded by the word "order". The names of its physical members (mononegaviruses/mononegavirads) are to be written in lower case, are not italicized and used without articles.

Order Inclusion Criteria

A virus is a member of the order *Mononegavirales* if

- its genome is: a linear, nonsegmented, single-stranded, non-infectious RNA of negative polarity; possesses inverse-complementary 3' and 5' termini; and does not possess a 5' cap, is not polyadenylated, and is not covalently linked to a protein
- its genome has the characteristic gene order 3'-UTR-core protein genes-envelope protein genes-RNA-dependent RNA polymerase gene-5'-UTR
- it produces 5-10 distinct mRNAs from its genome via polar sequential transcription from a single promoter located at the 3' end of the genome
- it replicates by synthesizing complete antigenomes
- it forms infectious helical ribonucleocapsids as the templates for the synthesis of mRNAs, antigenomes, and genomes
- it encodes an RNA-dependent RNA polymerase (RdRp) that is highly homologous to those of other mononegaviruses
- it forms enveloped virions with a molecular mass of 300–1,000 x 10^6; an S_{20W} of 550–>1,045; and a buoyant density in CsCl of 1.18–1.22 g/cm^3

Life Cycle

The mononegavirus life cycle begins with virion attachment to specific cell-surface receptors, followed by fusion of the virion envelope with cellular membranes and the concomitant release of the virus nucleocapsid into the cytosol. The virus RdRp partially uncoats the nucleocapsid and transcribes the genes into positive-stranded mRNAs, which are then translated into structural and nonstructural proteins. Mononegavirus RdRps bind to a single promoter located at the 3' end of the genome. Transcription either terminates after a gene or continues to the next gene downstream. This means that genes close to the 3' end of the genome are transcribed in the greatest abundance, whereas those toward the 5' end are least likely to be transcribed. The gene order is therefore a simple but effective form of transcriptional regulation. The most abundant protein produced is the nucleoprotein, whose concentration in the cell determines when the RdRp switches from gene transcription to genome replication. Replication results in full-length, positive-stranded antigenomes that are in turn transcribed into negative-stranded virus progeny genome copies. Newly synthesized structural proteins and genomes self-assemble and accumulate near the inside of the cell membrane. Virions bud off from the cell, gaining their envelopes from the cellular membrane they bud from. The mature progeny particles then infect other cells to repeat the cycle.

Paleovirology

Mononegaviruses have a history that dates back several tens of million of years. Mononegavirus "fossils" have been discovered in the form of mononegavirus genes or gene fragments integrated into mammalian genomes. For instance, bornavirus gene "fossils" have been detected in the genomes of bats, fish, hyraxes, marsupials, primates, rodents, ruminants, and elephants. Filovirus gene "fossils" have been detected in the genomes of bats, rodents, shrews, tenrecs, and marsupials. A Midway virus "fossil" was found in the genome of zebrafish.. Finally, rhabdovirus "fossils" were found in the genomes of mosquitoes and ticks.

Ranavirus

Ranavirus is one of five genera of viruses within the family Iridoviridae, one of the five families of nucleocytoplasmic large DNA viruses . *Ranavirus* is the only genus within *Iridoviridae* that includes viruses that are infectious to amphibians and reptiles, and one of only three genera within this family which infect teleost fishes, along with Lymphocystivirus and Megalocytivirus. The ranaviruses, like the megalocytiviruses, are an emerging group of closely related dsDNA viruses which cause systemic infections in a wide variety of wild and cultured fresh and saltwater fishes. As with megalocytiviruses, *Ranavirus* outbreaks are therefore of considerable economic importance in aquaculture, as epizootics can result in moderate fish loss or mass mortality events of cultured fishes. Unlike megalocytiviruses, however, *Ranavirus* infections in amphibians have been implicated as a contributing factor in the global decline of amphibian populations. The impact of ranaviruses on amphibian populations has been compared to the chytrid fungus *Batrachochytrium dendrobatidis*, the causative agent of chytridomycosis.

Etymology

Rana is derived from the Latin for "frog", reflecting the first isolation of a *Ranavirus* in 1960s from *Lithobates pipiens*, formerly *Rana pipiens*.

Taxonomy

The family *Iridoviridae* is divided into five genera which include *Chloriridovirus*, *Iridovirus*, *Lymphocystivirus*, *Megalocytivirus*, and *Ranavirus*. The genus *Ranavirus* is composed of at least 6 recognized viral species, 3 of which are known to infect amphibians (Ambystoma tigrinum virus (ATV), Bohle iridovirus (BIV), and frog virus 3).

Structure

Ranaviruses are large icosahedral DNA viruses measuring approximately 150 nm in diameter with a large single linear dsDNA genome of roughly 105 kbp which codes for around 100 gene products. The main structural component of the protein capsid is the major capsid protein (MCP).

Replication

Ranaviral replication is well-studied using the type species for the genus, frog virus 3 (FV3). Replication of FV3 occurs between 12 and 32 degrees Celsius. Ranaviruses enter the host cell by receptor-mediated endocytosis. Viral particles are uncoated and subsequently move into the cell nucleus, where viral DNA replication begins via a virally encoded DNA polymerase. Viral DNA then abandons the cell nucleus and begins the second stage of DNA replication in the cytoplasm, ultimately forming DNA concatemers. The viral DNA is then packaged via a headful mechanism into infectious virions. The *ranavirus* genome, like other iridoviral genomes is circularly permuted and exhibits terminally redundant DNA.

Transmission

Transmission of ranaviruses is thought to occur by multiple routes, including contaminated soil, direct contact, waterborne exposure, and ingestion of infected tissues during predation, necrophagy or cannibalism. Ranaviruses are relatively stable in aquatic environments, persisting several week or longer outside a host organism.

Epizoology

Amphibian mass mortality events due to *Ranavirus* have been reported in Asia, Europe, North America, and South America. Ranaviruses have been isolated from wild populations of amphibians in Australia, but have not been associated with mass mortality on this continent.

Pathogenesis

Synthesis of viral proteins begins within hours of viral entry with necrosis or apoptosis occurring as early as a few hours post-infection.

Gross Pathology

Gross lesions associated with *Ranavirus* infection include erythema, generalized swelling, hemorrhage, limb swelling, and swollen and friable livers.

Renibacterium Salmoninarum

Renibacterium salmoninarum is a member of the Micrococcaceae family. It is a Gram-positive, intracellular bacterium that causes disease in young salmonid fish. The infection is most commonly known as Bacterial Kidney Disease but may also be referred to as BKD, White Boil Disease, Dee Disease, Salmonid Kidney Disease and Corynebacterial Kidney Disease. It is of significant ecologic importance due to its affect on both farmed and wild salmonids. The disease is found in North America, Europe, Japan, Chile and Scandinavia, and is spread both vertically and horizontally. Pacific salmon appear to be the most susceptible to the disease.

Clinical Signs

The severity of clinical signs is very variable. There may be no outward clinical signs, or fish may show signs of lethargy and anaemia. Haemorrhagic skin lesions and exophthalmos may develop.

On postmortem examination there are normally signs of necrosis and granulomatous inflammation on the internal organs, especially the kidney. A diagnosis cannot be made based on clinical signs, instead laboratory tests such as specialised bacterial culture, ELISA, PCR and fluorescent antibody testing are necessary to identify the bacteria. Ideally more than one test should be used to confirm diagnosis.

Treatment and Control

Oral or injectable antibiotics should be used to treat the infection. Intraperitoneal vaccination can also be used to treat fish in an outbreak. Prevention is very important, and husbandry measures such as segregation, culling and decontamination should be used to ensure infection is not introduced.

Snakehead Rhabdovirus

Snakehead rhabdovirus (SHRV) is a *Novirhabdovirus* that affects various species warm water wild and pond-cultured fish in Southeast Asia, including snakehead species for which it is named.

Isolation

Investigations were conducted in the 1970s and 1980s to search for infectious agents that cause or contribute to a severe ulcerative disease, known as epizootic ulcerative syndrome (EUS), which causes high mortalities among a variety of fish species in Asia and the Pacific region. SNRV was isolated by virologist W. Wattanavijarn from a snakehead murrel, *Channa striata* (formerly *Ophicepahalus striatus*), a species of snakehead fish, which exhibited signs of epizootic ulcerative syndrome following an outbreak of this syndrome in Thailand.

The new viral isolate was identified as a rhabdovirus based on images and measurements taken via transmission electron microscopy and biochemical test results.

Plaque neutralization tests and immunofluorescence tests demonstrated that the new isolate was serologically unrelated to several other fish rhabdoviruses, including anquilla rhabdovirus (EVX), infectious hematopoietic necrosis virus (IHNV), pike fry rhabdovirus (PFRV), spring viremia of carp virus (SVCV), and viral hemorrhagic septicemia virus (VHSV). Subsequent infection studies exposing healthy snakehead murrels from a susceptible population to SHNV did not develop clinical signs associated with EUS, suggesting that SHRV does not play a significant role in EUS.

Taxonomy

It is classified as a *Novirhabdovirus* because it possesses a nonvirion gene (NV), which is the distinguishing feature of this genus. The NV gene is situated between the glycoprotein (G) and the polymerase (L) genes, and contains a single open reading frame (ORF), which is 335 nucleotides in length. It is called the "nonvirion" gene because no corresponding protein is present in the virion. Although the NV genes of novirhabdoviruses are similar in size, including approximately 110-122 codons, little sequence homology is evident.

Coding sequences of its glycoprotein (G) genes were found to be similar to the three other presently-classified novirhabdoviruses, VHSV, IHNV, and *Hirame rhabdovirus* (HIRRV), with between 36% and 47% amino acid identity.

Structure

SHRV is an enveloped, bullet-shaped RNA virus measuring approximately 170 nm x 60 nm with an electron dense nucleocapsid.

Genome

The SHRV genome is a non-segmented, negative sense RNA genome that is approximately 11,550 nucleotides in length, with six ORFs, including the matrix (M) gene, the nonvirion (NV) gene, the nucleoprotein (N) gene, the phosphoprotein (P) gene, the polymerase (large protein, L) gene, and the viral glycoprotein gene (G), arranged in the following order: 3'-N-P-M-G-NV-L-5'.

Replication

Viral replication occurs at both 15°C and 28°C in cell line derived from snakehead and carp, although the optimal temperature range for viral replication is between 28°C and 31°C.

Stability

SHRV has been demonstrated to become inactivated following treatment with acid (pH = 3), chloroform (50%), and heat (56°C). SHRV in distilled water can be completely inactivated by less than five minutes of exposure to 12.5 ppm chlorine, 50 ppm iodine, or a 1:2000 dilution of peroxygen disinfectant.

Exposure of infective virions in cell culture material to 2% formalin reduced infectivity by 99.9% after 5 minutes, and completely after 30 minutes. However, exposure of infectious cell culture material to 0.025% formalin for 60 minutes caused only a negligible reduction in infectivity, and more than 50 ppm chlorine was needed to inactivate the virus Also in cell culture fluids, exposure to 500 ppm iodine for 30 minutes did not reduce infectivity. The virus does not lose infectivity when exposed in cell culture fluids to malachite green for 6 hours at 5 ppm.

Experimental Infections

Snakehead Infections: Infection studies with SHNV did not produce disease in exposed healthy snakehead murrels.

Zebrafish Infections

The first experimental infections conducted with SHRV in zebrafish demonstrated that the NV gene played no important role in the pathogenesis of infection. Subsequent experimental infections exposed zebrafish embryos, juveniles, and adults to SHRV by immersion and/or intraperitoneal (IP) injection.

Whereas embryos and larvae were susceptible to infection by immersion, adult zebrafish were only susceptible to infection by IP injection. Histopathology of infected embryos and juvenile fish revealed vascular monocyte accumulation, accumulation of cellular debris in the gas bladder, necrosis of hepatocytes, and necrosis of pharyngeal epithelial cells. SHRV induced interferon (IFN) and orthomyxovirus resistance (Mx) gene expression in zebrafish at levels dependent on route of inoculation and fish age.

Spironucleus Salmonicida

Spironucleus salmonicida is a species of fish parasite. The species creates foul-smelling, pus-filled abscesses in muscles and internal organs of aquarium fish. In the late 1980s when the disease was first reported, it was believed to be caused by *Spironucleus barkhanus*. Anders Jørgensen was the person that found out what species really caused the disease.

Spring Viraemia of Carp

Spring viraemia of carp, also known as Swim Bladder Inflammation, is caused by a rhabdovirus called Rhabdovirus carpio.

It is the cause of contagious infection seen in all species of carp around 1 or 2 years of age in the Spring. The disease is widespread accross the American and European continents.

Transmission is mainly horizontal via the faeces, but vectors such as the leech and crustaceans may also spread the disease.

Clinical Signs and Diagnosis

The virus enters through the gills and infects the swim bladder, leading to inflammation and buoyancy problems. Systemic signs may include skin changes, decreased feeding and death.

Diagnosis relies on virus isolation, immunofluorescence, ELISA or PCR. Electron microscopy may reveal virus particles in the swim bladder wall.

Treatment and Control

No treatment exists for the disease. Biosecurity measures such as prompt detection, isolation and disinfection of tanks and equipment is important to prevent the spread of the disease.

Trypanosoma

Trypanosoma is a genus of kinetoplastids (class Kinetoplastida), a monophyletic group of unicellular parasitic flagellate protozoa. The name is derived from the Greek *trypano* (borer) and *soma* (body) because of their corkscrew-like motion. All trypanosomes are heteroxenous (requiring more than one obligatory host to complete life cycle) and are transmitted via a vector. The majority of species are transmitted by blood-feeding invertebrates, but there are different mechanisms among the varying species. Then in the invertebrate host they are generally found in the intestine and normally occupy the bloodstream or an intracellular environment in the mammalian host.

Trypanosomes infect a variety of hosts and cause various diseases, including the fatal human diseases sleeping sickness, caused by *Trypanosoma brucei*, and Chagas disease, caused by *Trypanosoma cruzi*.

The mitochondrial genome of the *Trypanosoma*, as well as of other kinetoplastids, known as the kinetoplast, is made up of a highly complex series of catenated circles and minicircles and requires a cohort of proteins for organisation during cell division.

Selected Species

Species of *Trypanosoma* include the following:

- *T. ambystomae* in amphibians
- *T. antiquus* Extinct (Fossil in Eocene amber)
- *T. avium*, which causes trypanosomiasis in birds
- *T. boissoni*, in elasmobranch
- *T. brucei*, which causes sleeping sickness in humans and nagana in cattle
- *T. cruzi*, which causes Chagas disease in humans
- *T. congolense*, which causes nagana in ruminant livestock, horses and a wide range of wildlife
- *T. equinum*, in South American horses, transmitted via Tabanidae,

- *T. equiperdum*, which causes dourine or covering sickness in horses and other Equidae, it can be spread through coitus.
- *T. evansi*, which causes one form of the disease surra in certain animals (a single case report of human infection in 2005 in India was successfully treated with suramin)
- *T. everetti*, in birds
- *T. hosei* in amphibians
- *T. levisi*, in rats
- *T. melophagium*, in sheep, transmitted via *Melophagus ovinus*
- *T. paddae*, in birds
- *T. parroti*, in amphibians
- *T. percae*, in the species *Perca fluviatilis*
- *T. rangeli*, believed to be nonpathogenic to humans
- *T. rotatorium*, in amphibians
- *T. rugosae*, in amphibians
- *T. sergenti*, in amphibians
- *T. simiae*, which causes nagana in pigs. Its main reservoirs are warthogs and bush pigs
- *T. sinipercae*, in fishes
- *T. suis*, which causes a different form of surra
- *T. theileri*, a large trypanosome infecting ruminants
- *T. triglae*, in marine teleosts
- *T. vivax*, which causes the disease nagana, mainly in West Africa, although it has spread to South America

Hosts, Life Cycle and Morphologies

Two different types of trypanosomes exist, and their life cycles are different, the salivarian species and the stercorarian species.

Stercorarian trypanosomes infect the insect, most often the triatomid kissing bug, develop in its posterior gut and infective organisms are released in the faeces and deposited on the skin of the host. The organism then penetrates and can disseminate throughout the body. Insects become infected when taking a blood meal. Salivarian trypanosomes develop in the anterior gut of insects, most importantly the Tsetse fly, and infective organisms are innoculated into the host by the insect bite before it feeds.

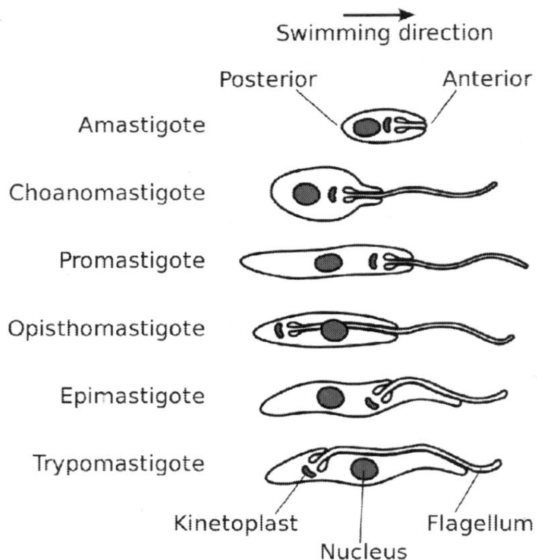

Figure: *The six main morphologies of trypanosomatids.*

As trypanosomes progress through their life cycle they undergo a series of morphological changes as is typical of trypanosomatids. The life cycle often consists of the trypomastigote form in the vertibrate host and the trypomastigote or promastigote form in the gut of the invertebrate host. Intracellular lifecycle stages are normally found in the amastigote form. The trypomastigote morphology is unique to species in the genus *Trypanosoma*.

Vibrio Anguillarum

Vibrio anguillarum is a Gram negative, curved rod bacterium with one polar flagellum. It is an important pathogen of cultured salmonid fish, and causes the disease known as vibriosis or red pest of eels. The disease has been observed in the salmon, the bream, the eel, the mullet, the catfish and Tilapia amongst others.

The organism is most prevalent in late summer in salt or brackish water and transmission is mainly horizontal by direct contact. It is widely distributed across the world.

Clinical Signs and Diagnosis

There are multiple haemorrhages in the body and skin changes signifying systemic involvement. Splenomegaly may be evident in young fish.

Diagnosis relies on culture of *V. anguillarum* and the use of monoclonal antibodies.

Treatment and Control

Various antibiotics such as ampicillin, chloramphenicol, nalidixic acid derivatives, nitrofurans, sulphonamides and trimethoprim can be used to treat the fish. Resistance is emerging however.

A vaccine against *V. anguillarum* is available.

Aeromonas (Hole in the Side Disease)

This disease is a severe gram-negative bacterial infection that starts out as small red spots, and soon develops into large ulceration's that eat their way through the fish. As the severity increases and the fish's immune system is being destroyed, multiple symptoms will start to appear. One of the diseases that is caused by Aeromonas bacteria is: Bacterial Hemorrhagic Septicemia (red streaking in fins and body, swelling, bloat, loss of scales, skin eaten away, abscessed ulceration's etc.) There are several different forms of Aeromonas bacteria. Including viral strains of Aeromonas that cannot be cured with the medications we have today.

We like to use a combination of Oxytetracycline mixed into the feed, and Forma-Green in the water to prevent a secondary fungal infection from developing on the ulceration's. Many people cure the Aeromonas infection with the Oxytetracycline and then find after the completion of the treatment, the sores are still open and red. This is due to saprolegnia fungus that has invaded the ulceration's and is acting like a parasite, eating away at the sores. This is why it is important to use an anti-fungal medication in the water along with the Oxytetracycline in the feed. We like the Forma-Green the best, but you can use a 0.75% Malachite Green solution instead.

If you can separate the infected fish early on and treat them in isolation tanks, you can prevent your whole pond from being infected in most cases. I like to bump up the temperature in the isolation tanks to 75-80° when possible. This will speed up the treatment. We have cured some severe cases of Aeromonas in two weeks with this method.

Using Antibiotics

There is much controversy when it comes to using antibiotics on tropical fish and Koi. Applications are confusing, and everywhere you go for advice, someone is telling you something different.

To clear a few things up, we have created this section, so treatments with antibiotics are less confusing. If you have purchased our products here at National Fish Pharmaceuticals, you will notice on the label that we tell you to use the product every 24 hours with a 25% water change before each treatment. Then we instruct you to treat for 10 days.

However, if you have a body of water that is 100 gallons of water or larger, the antibiotics will stay effective much longer, and not dissipate out as quickly as say... a 10 gallon tank. So instead of treating every 24 hours, we suggest treating every 3 days with a 25% water change before each treatment. This will not only save you some water changes, but some money on medications also.

Rules of thumb:

- Always use antibiotics for at least 10 days to prevent a resistant strain of bacteria from developing.
- Antibiotics may not be used as preventatives, or as a "dip". You can create problems doing this.
- Never mix any antibiotics together without proper consultation. Some antibiotics are not compatible when mixed, and the results can be fatal to your fish.
- If you are treating your water with antibiotics, do not feed your fish during the treatment time. This may cause Ammonia levels to become toxic, and besides...fish can go up to 30 days without eating anything and be perfectly fine.

National Fish Pharmaceuticals Tips and Prevention

Here you will find many ways to improve the quality of the environment for your tropical fish. And have a better understanding of fish keeping in general. We want to help make your hobby or business fun, not frustrating!

Below are links to few topics that make for a good foundation.

- Water Quality
- Feeding Your Fish
- Filtration
- Medicating Your Fish

Pond and Aquarium Water Quality Tips

Many of the fish at your local pet stores come from different parts of the world, and therefore must be kept in the ideal environment to ensure success.

Make sure you have consulted with your local professional or pet shop owner before purchasing any tropical fish. Many fish such as Discus, require additives to their water to survive and remain healthy. The same goes for Koi, African Cichlids, Guppies, Livebearers, Gouramies, Tetras etc. Fish from different continents require different additives to re-produce the water conditions from the area where they came from. Most importantly, if you are using tap water, you need to get rid of the Chlorine and Chloramines. You also need to establish some nitrifying Bacteria, we suggest using Aqua Gold.

First of all, we recommend doubling up the filtration on your aquarium. So, if you have a 100 gallon tank, put enough filtration on it for a 200 gallon tank. This will cut down on water changes and ensure that you have enough biological filtration to handle the load of fish in your aquarium or pond.

Water Changes

You have to remember that a fish is swimming around in his own toilet, so good clean fresh water is important. On a double filtered tank, a 25% water change per week is sufficient. Usually, on most aquariums 25% per week is acceptable. The size of the water change depends on how many fish are in the tank, and also how much filtration you are using. Your water should be crystal clear, and you should not have any black or red algae in your tank. If this is the situation, you need to treat the tank or pond with Erythromycin.

Too many water changes can be bad because you will not give the biological bacteria enough time to become established. If you are keeping your aquarium or pond too clean, you are doing more harm than good. Not to mention the fact that if you are in there scrubbing, moving rocks around and constantly disturbing the tank, you will stress out the fish and cause them to become ill and eventually die.

Rule of Thumb: Let Nature Take it's Course!

Want to see your Koi or Goldfish breed in your pond? Try letting it get pea soup green during the summer and leave it that way. Come springtime you may have quite a surprise: hundreds of new baby fish!

- Water Quality
- Feeding Your Fish
- Filtration
- Medicating Your Fish

9

Aquarium Fish Diseases and Treatments

There are literally hundreds of afflictions that can effect the health of your fish. The most common maladies seen in home aquaria are usually either bacterial or parasitic in origin. Fungal infections are also sometimes seen, and occasionally physical ailments.

***Figure:** Fish Diseases*

Luckily, most fish ailments are easily diagnosed and can be treated with success. The most common of these afflictions are included here. How to prevent fish disease has steps you can take to reduce the possibility of disease and help to keep disease from spreading if it should occure. A table of contents is provided along with a diagnostic chart with links to appropriate medications.

Understanding how an aquarium and its filtration work to support aquatic life is vital in preventing fish ailments. The basics of life support are the same whether you have a freshwater aquarium, saltwater aquarium, or a mini reef.

Types of Fish Disease: Fish ailments can be separated into 4 general types including bacterial infections, fungal infections, parasitic or protozoan infections, and physical ailments and wounds.

- *Bacterial Diseases:* Bacterial diseases are usually characterized by red streaks or spots and/or swelling of the abdomen or eye. These are best treated by antibiotics such as penicillin, amoxicillin, or erythromycin.
- *Fungal Diseases:* Comon fungal infections often look like gray or white fluffy patches. are
- *Parasitic Diseases:* The most common parasitic disease called "Ich" can be treated most effectively with copper or malachite green in the right dosage. Most treatments will have copper as an ingredient. Many water treatments like "Aquari-Sol" will also contain copper as an ingredient. If the treatment you use is an anti-biotic or copper based, remember to remove all carbon from the filtration system.
- *Physical Ailments:* Physical Ailments are often the result of the environment. Poor quality water conditions can lead to fish gasping, not eating, jumping out of the tank, and more. Tank mate problems can result in nipped fins and bite wounds.

How to Prevent Fish Diseases: Some steps can be taken to reduce the possibility of your fish getting a disease. Following these precautions can also help keep fish diseases from spreading if they do occur.

- Buy only good-quality, compatible fish.
- Quarantine new fish before adding them to the aquarium. (A hospital tank can be used for this).
- Avoid stressing the fish with rough handling, sudden changes in conditions, or "bully" tankmates.
- Don't overfeed your fish.
- Remove sick fish to a hospital tank for treatment.
- Disinfect nets used to move sick fish.
- Don't transfer water from the quarantine tank to the main aquarium.

- Don't let any metal come in contact with the aquarium water.
- *Anti-biotics:* When using any anti-biotic, make sure the biological filtration in your aquarium is not destroyed. You want to be certain the treatment does not kill the nitrifying bacteria in your system at the same time it attacks harmful bacteria on your fish. Although most of the treatments available at the store state that they will not harm your biological filter, sometimes they wil. It is best to either monitor your ammonia and nitrite levels, or use an ammonia remover such as "AmQuel" to be sure your levels of ammonia don't become a problem.
- *Copper Treatments:* When using any medication which has copper as an ingredient, be aware that most plants will not do as well. Invertebrates, such as snails, can also be killed if the amount of copper is sufficient. Indeed, most snail removers are copper based.

Diagnostic Chart

If you notice something is wrong with your fish, a proper diagnoses is usually all you need to worry about. In most cases, a proprietary treatment (fish medications) purchased at a pet store will work very well.

Steps to diagnosing and treating your fish:

1. *Symptom:* Match the symptoms your fish is showing with those listed under the 'Symptom' column.
2. *Possible Cause:* Learn about each disease by clicking on the link under 'Possible Cause'.
3. *Medication:* Find the treatment for the disease under the 'Medication' column.
4. *Product Link:* Use the 'Compare Prices' column to compare similar products and merchant's prices.

Bacterial Diseases

Red Pest

Symptoms: Bloody streaks on fins or body. Red Pest is called such because of bloody streaks that appear on the body, fins and/or tail. These streaks could proceed to ulcerations and possibly lead to fin and tail rot with, in severe cases, the tail and/or fins falling off. As the disease is internal, external treatments are usually not effective, except in very slight cases. In slight cases, treat the aquarium with a disinfectant and clean the aquarium as best as possible. Do not feed a lot while the aquarium is being treated. To disinfect, use acriflavine (trypaflavine) or monacrin (monoaminoacridine) using a 0.2% solution at the rate of 1 ml per liter. Both disinfectants will color the water, but the color disappears as the disinfectants dissipates. If the fish do not appear to respond favorably, discontinue disinfections.

Then add an antibiotic to the food. With flake food, use about 1% of antibiotic and carefully mix it in. If you keep the fish hungry they should eagerly eat the mixture before the antibiotic dissipates. Antibiotics usually come in 250 mg capsules. If added to 25 grams of flake food, one capsule should be enough to treat dozens of fish. A good antibiotic is chloromycetin (chloramphenicol). Or use tetracycline. If you feed your fish frozen foods or chopped foods, try to use the same ratio with mixing. As a last resort add at most 10 mg per liter of water.

Mouth Fungus

Symptoms: White cottony patches around the mouth. Mouth Fungus is so called because it looks like a fungus attack of the mouth. It is actually caused from the bacterium Chondrococcus columnaris. It shows up first as a gray or white line around the lips and later as short tufts sprouting from the mouth like fungus. The toxins produced and the inability to eat will be fatal unless treated at an early stage. This bacteria is often accompanied by a second infection of an Aeromonas bacteria.

Penicillin at 10,000 units per liter is a very effective treatment. Treat with a second dose in two days. Or use chloromycetin, 10 to 20 mg per liter, with a second dose in two days. Other antibiotics can also be effective. Kanacyn (kanamycin) will treat both bacteria at once. Maracyn (erythromycin) is effective against C. columnaris, and using Maracyn 2 (minocycline) in conjuntion with it will treat the Aeromonas bacteria as well.

Tuberculos – Mycobacteriosis

Syn: fish tuberculosis, piscine tuberculosis, acid-fast disease, granuloma disease.

Symptoms: Emaciation, hollow belly, possibly sores.

Tuberculosis is caused by the bacterium Mycobacterium piscium. Fish infected with tuberculosis may become lethargic, hollow bellied, pale, show skin ulcers and frayed fins, have fin and scale loss, and loss of appetite. Yellowish or darker nodules may appear on the eyes or body and may deform the fish.

The main causes for this disease appears to be over crowding in unkempt conditions; ie. poor water quality. All fish species could be susceptible though some are more susceptible than others. Those most susceptible are the labyrinth air breathers like the Gouramis, Bettas, and Paradise Fish. Others include Neon Tetras, Discus, and the Ram Cichlid.

- There is no absolute treatment. However the most effective treatment known for this disease is to treat with Kanamycin and Vitamin B-6 for 30 days. Kanamycin can be purchased at your local fish store. Liquid baby vitamins work well as s Vitamin B-6 source. They are available at your local pharmacy. Add one drop per every 5 gallons of aquarium water during treatment.
- If the treatment is ineffective, the best thing to do is destroy the infected fish.
- If either unkempt conditions or over crowding are the suspected cause, correct the condition.

It is possible for humans to contract this disease so we recommend using caution when dealing with it. Humans are very rarely are at risk from aquariums though. It is more common to contract this disease from public swimming areas or as a food contaminant.

Dropsy

Symptoms: Bloating of the body, protruding scales. Dropsy is caused from a bacterial infection of the kidneys, causing fluid accumulation or renal failure. The fluids in the body build up and cause the fish to bloat up and the scales to protrude. It appears to only cause trouble in weakened fish and possibly from unkempt aquarium conditions. An effective treatment is to add an antibiotic to the food. With flake food, use about 1% of antibiotic and carefully mix it in. If you keep the fish hungry they should eagerly eat the mixture before the antibiotic dissipates. Antibiotics usually come in 250 mg capsules. If added to 25 grams of flake food, one capsule should be enough to treat dozens of fish. A good antibiotic is chloromycetin (chloramphenicol). Or use tetracycline. If you feed your fish frozen foods or chopped foods, try to use the same ratio with mixing. As a last resort add at most 10 mg per liter of water. Also, if unkempt conditions are the suspected cause, correct it.

Scale Protrusion

Symptoms: Protruding scales without body bloat. Scale protrusion is essentially a bacterial infection of the scales and/or body. A variety of bacterium could be the culprit here, as can unkempt aquarium conditions.

An effective treatment is to add an antibiotic to the food. With flake food, use about 1% of antibiotic and carefully mix it in. If you keep the fish hungry they should eagerly eat the mixture before the antibiotic dissipates. Antibiotics usually come in 250 mg capsules. If added to 25 grams of flake food, one capsule should be enough to treat dozens of fish. A good antibiotic is chloromycetin (chloramphenicol). Or use tetracycline. If you feed your fish frozen foods or chopped foods, try to use the same ratio with mixing. As a last resort add at most 10 mg per liter of water. Also, if unkempt conditions are the suspected cause, correct it.

Tail Rot and Fin Rot

Symptoms: Disintegrating fins that may be reduced to stumps, exposed fin rays, blood on edges of fins, reddened areas at base of fins, skin ulcers with gray or red margins, cloudy eyes.

Tail and fin rot appears to be a bacterial infection of the tail and/ or fins and may be caused by generally poor conditions, bully, or fin nipping tankmates. If aquarium conditions are not good an infection can be caused from a simple injury to the fins/tail. Tuberculosis can lead to tail and fin rot. Basically, the tail and/or fins become frayed or lose color. Over time the affected area slowly breaks down.

First, attempt to ascertain the cause. Then treat accordingly. Also, treat the water or fish with antibiotics. If added to the water, use 20 - 30 mg per liter. If the fish is to be treated add an antibiotic to the food. With flake food, use about 1% of antibiotic and carefully mix it in. If you keep the fish hungry they should eagerly eat the mixture before the antibiotic dissipates. Antibiotics usually come in 250 mg capsules. If added to 25 grams of flake food, one capsule should be enough to treat dozens of fish. A good antibiotic is chloromycetin (chloramphenicol) or tetracycline. If you feed your fish frozen foods or chopped foods, try to use the same ratio with mixing. As a last resort add at most 10 mg per liter of water. Also, if unkempt conditions are the suspected cause, correct it.

Fish Vibriosis

Symptoms: Lethargy, increased respiration, loss of appetite, skin hemorrhages, and death.

Vibrio is a genus of Gram-negative bacteria found primarily in saltwater or brackish water, and consisting of 70 or more strains. Fish Vibriosis involves a variety of infectious strains of Vibrio bacteria, most notably Vibrio anguillarum, V. ordalii, V. damsela, and V. salmonicida.

Fish Vibriosis occurs most often in marine animals or brackish water fish, though it can occasionally be found in tropical species. Fish contract the bacteria through open sores or feeding on dead fish that died from the disease. Hemorrhaging starts with reddening or blood streaks under the skin surface, becoming red spots on the ventral and lateral areas of the fish. Swollen dark lesions develop, turning into ulcers and release bloody pus. There may also be eye problems with cloudy eye, which can lead to pop-eye and eye loss.

The course of a vibriosis infection in fish is usually very rapid. Most infected fish die without showing more visual signs than the ulcers, and sometimes death may occur suddenly before any signs are noticed at all.

The best treatment includes oral antibiotics. Kanamycin is one of the best, also chloramphenicol or furazolidone are good. When treating with antibiotics, it must be done in a quarantine tank rather than the main aquarium. This is because antibiotics will damage the biological filter in the main tank, throwing the nitrification cycle into reverse and cause a spike in nitrites and ammonia after just a few days.

A word of caution: People can become infected by Vibrio bacteria when handling infected fish. Hence the nicknames of "fish handler's disease" and "aquarium handler's disease". People can become infected when water containing the bacteria comes into contact with cuts or open sores on the skin. The bacteria may be in swimming pools, aquariums, or coastal waters.

Vibrio infections are vary rare in the United States. Cholera is probably the most well-known illness caused by one of the vibrio bacteria. People sick with cholera usually have intestinal symptoms such as diarrhea and vomiting. Those with compromised immune systems can be at higher risk, even death, with some strains of Vibrio bacteria.

Protozoan Diseases

Velvet or Rust

Symptoms: Clamped fins, respiratory distress (breathing hard), yellow to light brown "dust" on body. This disease has the appearance of a golden or brownish dust over the fins and body. The fish may show signs of irritation, like glancing off aquarium decor, shortage of breath (fish-wise), and clamping of the fins. The gills are usually the first thing affected. Velvet affects different species in different ways. Danios seem to be the most susceptible, but often show no discomfort. This disease is highly contagious and fatal.

The best treatment is with copper at 0.2 mg per liter (0.2 ppm) to be repeated once in a few days if necessary. Acriflavine (trypaflavine) may be used instead at 0.2% solution (1 ml per liter). As acriflavine can possibly sterilize fish and copper can lead to poisoning, the water should be gradually changed after a cure has been effected.

Marine Velvet (*Amyloodinium ocellatum*)

Symptoms: Respiratory distress (fast breathing - gills opening more than 80 times per minute); White, yellow to light brown, or grey "dusty" appearance on body, Loss of appetite, Rubbing or scratching against decor or substrate.

Marine velvet is one of the most common maladies experienced in the marine aquarium, with the other being Marine Ich. It is found in all the oceans of the world and often infects wild and newly caught marine fish. It is a fast moving disease that can cause mass casualties. Primarily it infects the gills of fish but can attach itself to the body as well, burrowing deep into the skin's subcutaneous layer. Deaths are generally a result of interference to the respiratory system. This disease is highly contagious and fatal.

Chemical treatments for this disease include using copper. Follow the instructions provided by the manufacturer. Natural methods include hyposalinity, a quarantine tank with a low salinity. A danger with with using low salinity is in re-acclimating the fish to a higher salinity. You must be able to accurately measure the salinity and must increase it very slowly.

Costia

Symptoms: Milky cloudiness on skin.This is a rare protozoan disease that causes a cloudiness of the skin. The best treatment is with copper at 0.2 mg per liter (0.2 ppm) to be repeated once in a few days if necessary. Acriflavine (trypaflavine) may be used instead at 0.2% solution (1 ml per liter). As acriflavine can possibly sterilize fish and copper can lead to poisoning, the water should be gradually changed after a cure has been effected. Raising the water temperature to 80° - 83° F for a few days has also been effective.

Hexamita

Symptoms: The first symptom of slimy, white mucous feces, even while still eating and acting normal. Further signs are the fish hiding in the corner it's head down, head above the eyes gets thin, they blacken in color, and swim backwards. Hexamita are intestinal flagellated protozoa that attack the lower intestine. Discus and other large cichlids, especially Oscars, are especially prone to Hexamita. As it is a disease of the digestive tract, a wasting away or loss of appetite may be experienced.An effective treatment is the drug metronidazole. A combined treatment in the food (1% in any food the fish will eat) and in the water (12 mg per liter) is recommended. Repeat the water treatment every other day for three treatments.

(This disease is often confused with another disease called Head and Lateral Line Erosion (HLLE), which use to be called "hole-in-the-head" disease, because both these diseases are often seen simultaneously in the same fish. Head and Lateral Line Erosion disease looks like cavities or pits on the head and face. It is not a protozoan disease, but is actually caused by environmental conditions.)

Ich, Ick, White Spot Disease (*Ichthyophthirius multifiliis*)

Symptoms: Salt-like specks on the body/fins. Excessive slime. Problems breathing (ich invades the gills), clamped fins, loss of appetite.

Ich, Ick, or White Spot Disease is the most common malady experienced in the home aquarium. Luckily, this disease is also easily cured if caught in time! Ich is actually a protozoa called Ichthyophthirius multifiliis. There are three phases to the life cycle of this protozoa. Normally, to the amateur aquarist, the life cycle is of no importance. However, since Ich is susceptible to treatment at only one stage of the life cycle, an awareness of the life cycle is important.

- Adult phase - it is embedded in the skin or gills of the fish, causing irritation (with the fish showing signs of irritation) and the appearance of small white nodules. As the parasite grows it feeds on red blood cells and skin cells. After a few days it bores itself out of the fish and falls to the bottom of the aquarium.
- Cyst phase - after falling to the bottom, the adult parasite forms into a cyst with rapid cell divisions occurring.
- Free swimming phase - after the cyst phase, about 1000 free swimming young swim upwards looking for a host. If a host is not found within 2 to 3 days, the parasite dies. Once a host is found the whole cycle begins anew.

These three phases take about 4 weeks at 70° F but only 5 days at 80° F. For this reason it is recommended that the aquarium water be raised to about 80° for the duration of the treatment. If the fish can stand it, raise the temperature even higher up to 85°. The free swimming phase is the best time to treat with chemicals. Raising the aquarium temperature to 80° F will greatly shorten the time for the free swimming phase to occur. The drug of choice is quinine hydrochloride at 30 mg per liter (1 in 30,000). Quinine sulphate can be used if the hydrochloride is not available. The water may cloud but this will disappear. By reducing the time (with raised temperature) of the phases, you should be able to attack the free swimming phase effectively. Some aquarists like to use malachite green, but it tends to stain the plastic and silicone in the aquarium. Most commercial remedies contain malachite green and/or copper, which are both effective.

Marine Ich (Cryptocaryon irritans)

Symptoms: Salt-like specks on the body/fins. Rubbing or scratching against decor or substrate, Excessive slime. Problems breathing (ich invades the gills), Frayed fins, Loss of appetite, Cloudy eyes, Abnormal swimming. Marine ich or white spot disease is one of the most common maladies experienced in the marine aquarium, with the other being Marine Velvet. This protozoa has four phases to its life, lasting up to 38 days depending on the temperature of the environment. This parasite affects marine and brackish fish.

Aquarists are most familiar with the stage where the protozoa is infesting the host, the small white spots similar to a sprinkling of salt on the fish's body and fins. Unfortunately this visual clue is also the reason for difficulty in eradicating marine ich. Once the parasite has left the host's body many aquarists believe their fish is cured and the problem is solved and so they cease treatment, only to have another larger reoccurrence.

For eradication treatment must be carried through to completion, so understanding the parasite's life cycle will greatly increase your chances of success. The life cycle is outlined here:

- Trophont phase - when the parasite is growing in the skin or gills of the fish it appears as small white nodules, and the fish begins showing signs of irritation. It will spend 5 to 7 days (depending on the temperature) feeding on the fish. Once it reaches maturity it leaves the fish, reportedly after the lights go out. It is now called a protomont.
- Protomont phase - the protomont will free swim or will crawl about the substrate for several hours (2 to 18 hours) producing a sticky wall around itself with which it is able to adhere to a surface. Once it adheres it begins to turn into a cyst and is now called a tomont.

- Tomont phase - at this stage there is rapid cell divisions occurring, resulting in hundreds of daughter parasites that are called tomites. This stage can last anywhere from 3 to 28 days. Eventually the tomites hatch and begin swimming about looking for a host and are now called theronts.
- Theront phase - newly hatched, they are swimming about looking for a host which they must find within 24 hours or they will die. Once a host is found they turn into trophonts and the whole cycle begins anew.

The life cycle of this parasite can vary dramatically and is dependent on temperature, they cycle faster in a warmer environment. Ideally the parasite would be eliminated while on the host or shortly after leaving the host. However, those that are buried in the gills are immune to treatment until they leave the fish. This along with the variability of the cycle makes it difficult to treat in a timely manner.

So to rid the aquarium of this protozoa, it is recommended that you use a combination of water changes and chemical treatment, a multiple number of times.

- *Chemical:* Chemical treatments for this disease include using copper, formalin, or a combination of copper and formalin. Follow the instructions provided by the manufacturer.
- *Natural:* Natural methods include either a quarantine tank with a low salinity (hyposalinity) or large frequent water changes. For low salinity keep the specific gravity of the water at approximately 1.009-1.010 with temperatures of 78 - 80° F (25 - 27° C) for 14 days.

A danger with with using low salinity is in re-acclimating the fish to a higher salinity. You must be able to accurately measure the salinity and must increase it very slowly. For the water change method, replace 50% of the aquarium water daily for 14 days. This is perfectly safe method as long as temperature and salinity are the same, and this will remove the parasites while in a free swimming stage. Reportably some healthy fish can develop a limited immunity. This immunity is short-lived lasting only about six months and may not be a total immunity, being a small amount of infestation rather than extensive infestation.

Neon Tetra Disease

Symptoms: Whitened areas deep into the fishes' flesh. Muscle degeneration leading to abnormal swimming movements. So named for the fish it was first recognized on, the Neon Tetra. It is caused by the sporozoa Plistophora hyphessobryconis. Even though it is named after Neon Tetras, it can appear on other fish. Whitish patches appear as if just below the skin. In Neon Tetras it destroys the bright blue-green neon stripe. The organisms form cysts which burst and release spores. The spores penetrate further and form more cysts. Eventually, the spores migrate to the water and are eaten by other fish in the food. These spores migrate into the digestive tract, then the muscles, and a new infection starts. There is no known cure. It is best to destroy the infected fish and clean the aquarium.

Glugea and Henneguya

Symptoms: Similar to Lymphocystis, the fish will have nodular white swellings on fins or body. Glugea and Henneguya are sporozoans that form large cysts on the fish's body and release spores. Luckily, these diseases are very rare. The fish bloat up, with tumor like protrusions, and eventually die. No cure, as of yet. It is best to destroy the infected fish before the spores can spread.

Chilodonella

Symptoms: Dulling of the colors due to excessive slime, fraying of the fins, weakness, gill damage

This disease causes a blue white cloudiness on the skin and attacks the gills. Later the skin may be broken down and the gills destroyed. The fish may behave like they have irritations, by glancing off aquarium decor, they may have clamped fins and difficulty breathing.

Acriflavine (trypaflavine) may be used at 1% solution (5 ml per liter). As acriflavine can sterilize fish, the water should be gradually changed after a cure has been effected. It also helps to raise the temperature to about 80° F.

African Bloat or "Malawi Bloat"

Symptoms: The first sign of 'bloat' is loss of appetite which is then followed by swelling of the abdomen, labored breathing, listlessness, reclusiveness, possible red striations on the body, and stringy white feces.

There seems to be no explainable rationale as to its cause of bloat. Once a fish becomes afflicted it is often fatal. A fish that is not eating must be treated immediately or it can quickly become incurable and die. Though It is not certain what this disease is, it is generally believed to be caused by a protozoal parasite complicated by bacterial infection. Bloat is a serious malady often associated with African cichlids especially those from Lake Malawi, thus the common name 'Malawi Bloat'. The *Tropheus* species from Lake Tanganyika are also very susceptible.

The most common cause of this disease is stress and the first sign if illness is not eating. Stress can be caused by such things as transport, netting, poor water quality, insufficient diet, over feeding, and a lack of hiding places. Other causes, that are easily remedied, are an improper diet and adding too much salt to the water. Prevention is of utmost importance, and It is possibly to cure a fish if treated right away.

Following are some techniques aquarists use:

- Any new specimens you obtain can have bloat or will often soon develop it. When you first acquire them try to provide them with the same food that the dealer was feeding, and then wean them onto a good vegetable based diet; Spirulina flake and pellet.
- Some will soak the food in dissolved metronidazol and feed them that for the first few days when first obtained. Seachem makes a metronidazol that can be bound to food when used with their Focus product.
- A good vegetable based diet is important.
- A healthy group of fish will eat with gusto. But even though they can be very active feeders it is important to not overfeed them. Keep an eye on them, and if one is not eating with vigor some aquarists will then treat the tank with Clout.
- One author says that they will segregate an ailing fish the second they see signs of not eating, and then will do water changes every day for 5 days in the main aquarium.

Metronidazol is considered the most reliable cure and some use Clout as another cure, but do not use them together.

Parasitic Diseases

Argulus (Fish louse)

Symptoms: The fish scrapes itself against objects, clamped fins, visible parasites about 1/4 inch in diameter are visible on the body of the fish. The fish louse is a flattened mite-like crustacean about 5 mm long that attaches itself to the body of fish. They irritate the host fish which may have clamped fins, become restless, and may show inflamed areas where the lice have been.

With larger fish and light infestations, the lice can be picked off with a pair of forceps. Other cases can best be done with a 10 to 30 minute bath in 10 mg per liter of potassium permanganate. Or treat the whole tank with 2 mg per liter, but this method is messy and dyes the water.

Anchor Worm (Lernaea)

Symptoms: The fish scrapes itself against objects, whitish-green threads hang out of the fish's skin with an inflamed area at the point of attachment.

Anchor worms are actually crustaceans. The young are free swimming and borrow into the skin, go into the muscles and develop for several months before showing. They release eggs and die. The holes left behind are ugly and may become infected.

The anchor worm is too deeply imbedded to safely remove. Treatment can best be done with a 10 to 30 minute bath in 10 mg per liter of potassium permanganate. Or treat the whole tank with 2 mg per liter, but this method is messy and dyes the water.

Black Spot - Black Ick (diplopstomiasis)

Symptoms: The fish, very irritated, scrapes itself against objects, appears as small black specks or smudges on the body and around the mouth, and if heavily infected may experience blood loss.

Black Spot or Black Ick is rare in aquariums. It is generally seen in outdoor ponds, especially those with mud bottoms, but it can be introduced when adding new fish into the aquarium. Fish that are most readily susceptible are the Silver Dollar, Piranha, or other fish of these types. In general it does relatively little damage to the fish, even if they are heavily infested. The disease is caused by a parasite (larval trematode) that burrows into the skin of a fish where it forms a cyst that is about one millimeter in diameter. It has a complex lifecyle that in order to survive, requires fish eating birds or animals, snails, or fish at different stages infected with the disease. Black Spot is generally easy to cure. Either treat with salt baths or there are a number of commercially available treatments and preventatives.

Ergasilus

Symptoms: The fish scrapes itself against objects, whitish-green threads hang out of the fish's gills.This parasite is like the anchor worm, but is smaller and attacks the gills instead of the skin. Treatment can best be done with a 10 to 30 minute bath in 10 mg per liter of potassium permanganate. Or treat the whole tank with 2 mg per liter, but this method is messy and dyes the water.

Flukes

Symptoms: The fish scrapes itself against objects, rapid gill movement, mucus covering the gills or body, the gills or fins may be eaten away, the skin may become reddened.There are many species of flukes, which are flatworms about 1 mm long, and several symptoms that are visible. They infest gills and skin much like ich, but the difference can be seen with a hand lens. You should be able to see movement and possibly eye spots, which is not found in ich. Gill flukes will eventually destroy the gills thus killing the fish. Symptoms of a heavy infestations are pale fish with drooping fins, rapid respiration, glancing off aquarium decor, and /or hollow bellies.

Treatment can best be done with a 10 to 30 minute bath in 10 mg per liter of potassium permanganate. Or treat the whole tank with 2 mg per liter, but this method is messy and dyes the water.

Nematoda

Symptoms: Worms hanging from the anus.Nematodes (threadworms) infect just about anywhere in the body but only shows itself when they hang out of the anus. A heavy infestation causes hollow bellies. Lighter infestations usually cause no problems with the fish.

Short of destroying the fish, which is easier, two treatments have been suggested. First treatment; soak the food in parachlorometaxylenol and give the fish a bath or treat the aquarium with 10 ml per liter. The bath should last for several days. Second treatment; find special food containing thiabendazole as a nematode (threadworm) cure and hope the fish will eat it.

Leeches

Symptoms: Leeches are visible on the fish's skin. Leeches are external parasites and affix themselves on the body, fins, or gills of the fish. Usually they appear as heart shaped worms (they are just curled up) attached to the fish. They are usually introduced to the aquarium via plants or snails.

Since leeches are sucking and borrowing into the surface of the fish, removal with forceps can cause great damage, if not death, to the fish. If the fish is bathed in a 2.5 percent solution of salt for 15 minutes, most of the leeches should just fall off. Those that do not will be affected enough to remove with forceps with minimal damage. Another treatment is to add Trichlorofon at 0.25 mg/l to the aquarium. Live plants should be removed and treated with potassium permanganate at 5 mg/l before replanting.

Uronema Marinum

Symptoms: Skin scraping, pale discoloration, loss of color, weight loss, dehydration, flashing, and rapid breathing.The saltwater parasite, *Uronema marinum*, is a free-living ciliated protozoa that can cause fatal infections in marine fish. It is an opportunistic feeder that normally eats on bacteria, but when the immunization of a fish is low it will attack, invading the fish's muscles and internal organs. This infestation is often the result of the introduction of a new fish, overcrowding, and poor water quality resulting from a high organic load in the aquarium. This parasite is difficult to identify as the symptoms can also be indicative of other parasitic and bacterial problems. However, it can be debilitating and ultimately fatal to a variety of marine fish including Tangs, especially the Yellow Tang, Angelfish species especially those in the genus *Centropyges*, Seahorses, many species of Butterflyfish, yellow headed Jawfish, and others.

The best way to avoid the problem is to keep your current tank free from infestation. Quarantine all new fish for a period of three weeks, improve the water quality of the tank, and reduce the stress level in the aquarium by reducing the number of fish and incorporating places for fish to hide and rest.

There are several types of medications that can be used to treat infected fish:

- Medications such as Malachite green, Copper Sulfate, or Methylene blue. Use caution and be sure to follow the manufacturers instructions.
- Freshwater bath - place infected fish in the freshwater bath for a period of three minutes or until the fish shows signs of stress.
- Low salinity (hypo salinity) treatment - lower the salinity in the quarantine tank to a specific gravity of 1.011 and maintain at this salinity for 21 days. Do not use this treatment with invertebrates or especially sensitive fish such as sharks and rays.
- Nitrofurazone - an antibiotic that has some antiparasitic action, and can be helpful when used along with formalin dips.

Fungal Diseases

Fungus (Saprolegnia)

Symptoms: Tufts of dirty, cotton-like growth on the skin, can cover large areas of the fish, fish eggs turn white. Fungal attacks always follow some other health problem like parasitic attack, injury, or bacterial infection. The symptoms are a gray or whitish growth in and on the skin and/or fins of the fish. Eventually, if left untreated, these growths will become cottony looking. The fungus, if left untreated, will eventually eat away on the fish until it finally dies.

After ascertaining the initial cause of the fungus and remedying that, use a solution of phenoxethol at 1% in distilled water. Add 10 ml of this solution per liter of aquarium water. Repeat after a few days if needed, but only once more as three treatments could be dangerous to aquarium inhabitants. If the symptoms are severe the fish can be removed from the aquarium and swabbed with a cloth that has been treated with small amounts of povidone iodine or mercurochrome. For attacks on fish eggs, most breeders will use a solution of methylene blue adding 3 to 5 mg/l as a preventative measure after the eggs are laid.

Ichthyosporidium

Symptoms: Sluggishness, loss of balance, hollow belly, external cysts and sores.

Ichthyosporidium is a fungus, but it manifests itself internally. It primarily attacks the liver and kidneys, but it spreads everywhere else. The symptoms vary. The fish may become sluggish, lose balance, show hollow bellies, and eventually show external cysts or sores. By then it is usually too late for the fish. Treatment is difficult. Phenoxethol added to food as a 1% solution may be effective. Chloromycetin added to the food has also been effective. But both of these treatments, if not watched with caution, could pose a risk to your fish. It is best, if diagnosed soon enough, to destroy the affected fish before the disease can spread.

Miscellaneous

Head and Lateral Line Erosion Disease (HLLD or HLLE) - Hole-in-the-Head Disease

Symptoms: Begins as small pits on the head and face, usually just above the eye. If untreated, these turn into large cavities and then the disease progresses along the lateral line.

Head and Lateral Line Disease is also known as Hole-in-the-Head Disease, Lateral Line Erosion (LLE), and Lateral Line Disease (LLD). It is attributed to a nutritional deficiency of one or more of: Vitamin C, Vitamin D, calcium, and phosphorus. Though its cause is not definitively determined, it is thought to be caused by a poor diet or lack of variety, lack of partial water changes, or over filtration with chemical media such as activated carbon.

HLLE has been reversed by one or more of the following treatments:

- Increase frequent water changes.
- Add vitamins to frozen foods.
- Add the addition of flake foods, as they are enriched with vitamins.
- Add greens, either frozen or in leaf form, to the diet.
- Decrease the amount of beef heart as it lacks many critical nutrients.
- Remove activated carbon filtration.

(This disease is often confused with another disease called 'Hexamita', because both these diseases are often seen simultaneously in the same fish. Hexamita is a protozoan disease that attacks the lower intestine. Discus and other large cichlids, especially Oscars, are especially prone to Hexamita.)

Eye Problems

Symptoms: Cloudy cornea, opaque lens, pop eye, swelling, blindness.

- Cloudy cornea can result from a bacterial invasion. Antibiotics may help.
- Opaqueness can result from poor nutrition or a metacercaria invasion (grubs). Try foods with added vitamins and changing the diet to include variety.
- Pop eye (exophtalmia) can result from rough handling, gas embolism, tumors, bacterial infection, or vitamin A deficiency. Gas bubble or bacterial infection can be treated successfully with penicillin or amoxicillin.
- Blindness can be caused by poor nutrition or excessive light. Lowering the light level and a change of diet to include lots of variety may help prevent it.

Swim-bladder Disease

Symptoms: Abnormal swimming pattern, difficulty maintaining equilibrium. Swim bladder problems usually indicate another problem listed here. If you suspect swim-bladder problems in a fish, first check and treat it for other diseases as listed below:

- Congenitally deformed bladder
- Cancer or tuberculosis in organs adjacent to the swim bladder

- Constipation
- Poor nutrition
- Chilling or rapid fluctuations in temperature
- Serious parasitic infestation
- Serious bacterial infestation

If you have eliminated other causes, make sure you are feeding the right food and make sure the fish is not constipated. Give it live food for awhile to ensure it is getting enough roughage. Also, check the temperature for your fish's requirements and keep the temperature stable.

Non-infectious Maladies

Physical Injuries: Even in the best of aquariums under the supervision of the most astute aquarists, injuries occur. Some times a bully fish is the culprit, or sharp decor. Sometimes there appears to be no explanation. As in the human world, accidents happen.

If the cause of the injury is obvious, it should be remedied. Then the injury should be treated. The injury should be touched with 2% Mercurochrome, which is supplied commercially. Also, depending on the fish's tolerance to water conditions, keeping the fish in slightly acid water should speed recovery (pH 6.6). Minor injuries, if the water conditions are good, should just heal themselves.

Constipation

Some fish are more susceptible to constipation than others. Usually fish with more compressed bodies like angelfish and silver dollars. Symptoms are loss of appetite and swelling of the body. The cause is almost always diet.

Usually, with a change of diet, the condition rights itself. But in stubborn cases try dried food that has been soaked in medicinal paraffin oil. Glycerol or castor oil may also be used. If the diet is changed on a regular basis and live foods offered occasionally this condition may never occur.

Tumors

Tumors can be caused by a virus or a cancer, but most tumors are genetic. The genetic tumors may be caused from too much hybridization, common amongst professional breeders.

Practically all tumors are untreatable. If the fish is in distress, it should be destroyed.

Congenital Abnormalities

Abnormalities usually occur when professional breeders are trying to acquire certain strains in breeds. Most are beneficial abnormalities like albinism or extra finnage. But undesirable abnormalities crop up and are usually culled out by the breeder. However, such abnormalities sometimes happen in the amateur aquarium.

If the abnormality is not life threatening or degrades the quality of life, just leave it be and brag to your friends about the unusual inhabitant. Otherwise, the fish should be humanely destroyed.

Viral Diseases

Lymphocystis

Symptoms: Nodular white swellings (cauliflower) on fins or body.

Lymphocystis is a virus, and being a virus, it affects the cells of the fish. It usually manifests itself as abnormally large white lumps (cauliflower) on the fins or other parts of the body. It can be infectious, but is usually not fatal. Unfortunately there is no cure, but fortunately this is a rare disease. There are two suggested treatments. One treatment is to remove and destroy the infected fish as soon as possible. The other treatment is to simply separate the infected fish for several months and hope for remission, which usually does occur.

Unfortunately, even the most dedicated aquarist will from time to time have an outbreak of disease in the aquariums. How you handle these situations may be critical to your success as a hobbyist. One of the most common mistakes aquarists make is to medicate too soon. Very often the fish will fight off the disease on its own, provided that its immunity system is strong. I rarely medicate an aquarium. Rather, I rely on a series of partial water changes to rid the aquarium of parasites, pathogens, and other disease causing organisms. What follows are some of the diseases you are likely to run into and my recommended treatment. Please be aware that this is from my personal experience.

Koi, Pond Fish Disease

Disease is "any condition that deviates from a normal or healthy state". It might well be considered as a situation out of whack, out of balance. There are several types of and ways to describe disease: social, environmental, nutritional, genetic, infectious, parasitic... This article attempts to delineate some of the more common, easily diagnosed complaints of pond fish keeping, and offers some suggestions on prevention and treatment. As with humans, fish diseases are rarely caused by single events, nor are their treatments effected by one simple change or addition. Very often water quality and nutrition play a central role in "re-centering" a system, including the living and non-living portions. When your fish(es) seem out of sort, first look to the causes; typically water quality degradation, and rectify those shortcomings, before or rather than resorting to therapeutics.

Prevention

Prevention is always the best medicine. In an aquatic system, there are three general areas of disease prevention: environmental stress, fungus and bacteria, and parasites.

To overcome new-comer stress use a good conditioner like Novaqua when introducing new livestock; this will neutralize chlorine, detoxify some metals, and provide an extra slime coating for the fish. If a chlorine neutralizer is not wanted, Polyaqua may be used. To prevent fungus and bacteria infections, add about a tablespoon of salt per five gallons to the system. If live plants will be in the system, use only a tenth dose. To prevent fungus and bacteria in a system, use adequate filtration and frequent water changes. Ultraviolet sterilizers, ozone generators, and protein skimmers are high-technology add-ons that can boost water quality and extend safety margins from over-feeding and crowding.

For parasite control use quarantine, prophylactic dips and, if necessary copper. When using copper as a prevention use only a minimum dose, (0.15 ppm copper initial dose maximum)

Do not use copper with invertebrates. Use Tetra medicated food or make your own medicated food when introducing new fish or when invertebrates are present.

If fish still become sick, check for possible causes of stress and eliminate it if at possible. Identify the disease and treat accordingly. Carefully record what you observed and tested for, as well as what you treated with and how.

Stress

Any system's condition that is not good for the livestock may cause stress and too much stress usually leads to disease. The most common sources of stress are:

- Improper pH or drastic and\or sudden changes in pH.
- Improper temperature or sudden changes in temperature.

- mproper salinity or water density for extended periods.
- Improper hardness, or sudden change in hardness.
- Pounding on the system, or sudden movements that scare the fish.
- Aggressive pond-mates.
- Poor diet.
- High metabolite levels. (A high nitrite level prevents oxygen from reaching the cells and may cause suffocation or brain damage).
- High nitrate levels. (This may lower the PH as well).
- Any measurable ammonia level. (80% of all fish waste is in the form of ammonia and is extremely toxic if not converted immediately into nitrite).
- Other toxins. (Chlorine, copper, detergents, iron, lead, zinc, commercial ammonia, nicotine, perfume or cologne, oil, paint fumes, insecticides including contamination from dog and cat flea collars, etc.).
- Too little or too much dissolved gas in the water, or a rapid change from water that is saturated with gases to a "normal" gas saturation level. This can be a problem when releasing newly purchased fish into the system. If the water is not mixed, the fish may suffer a condition that is very similar to "the bends" and may be just as deadly.
- Too much or too little light. Too much light (no, or short periods of darkness), speeds up the metabolism of the fishes and does not allow the fish to rest. Too little light and the fish may be lethargic and not eat properly.
- Dirty or cloudy water. Cloudy water is usually caused by bacteria. The bacteria in the water uses all the available oxygen and the fish suffocate.
- No hiding places for the fish to feel safe.
- Loss of mucous of the fish's slime coating. This condition may be caused by stress instead of the other way around, but once the slime coating is lost, additional stress is incurred.
- Infectious (bacterial & fungal) and protozoan diseases.
- Any other sudden changes in the environment.
- Overcrowding.

Stress triggers the release of epinephrine, a hormone which prompts the fish to get ready to fight or flee. (The hormone used to be called adrenaline.) This increases the heart rate, blood pressure and respiration. At worst continual stress will cause a fish to die of exhaustion. At best the fish's immune system may become so over-stimulated that it functions improperly and the fish dies from a disease that would not kill an unstressed organism. A over-stressed fish usually becomes sick. If the stress is severe the fish may go into shock and die immediately. Most aquatic systems have a constant supply of fungus, bacteria, and parasites that have little or no effect on a healthy fish. Once a fish is under stress however, it may fall prey to the disease organisms that comes along or are already present but that it might have resisted without being over-stressed.

Fish and invertebrates are more dependent on their environment than any of the higher animals. They are totally dependent on their owners to provide them with proper living conditions.

Parasites Ich: White spot disease. Think of ich (Ichthyophthirius) as an army of individual animals living off the juices of the fish. While on the fish they are usually encysted and protected, and cannot be killed until they drop off the fish to reproduce on the tank bottom. Therefore treatment must continue through the entire life cycle of the parasite which usually is about 4 to 5 days. A low temperature can slow the cycle down to 6 or 7 days. A high temperature can speed it up to as little as 3 days. A high temperature (85 degrees or more) can kill parasites by itself but may also kill the fish or add to the stress of a fish with parasite choked gills.

Any parasitic infection is usually easy to cure if treated quickly with an effective dose of copper. If the dosage is too low, not all parasites are killed and re-infection results. If treatment is delayed, the parasites may become so numerous that they choke the gills and the fish suffocates or the fish becomes so weak it cannot recover. Treatment should continue for at least 4 days and a good rule of thumb is to treat the system every day until no sign of infection is visible, then treat one more day.

Each parasite leaves a wound where it was attached to the fish. These wounds are easily infected by bacteria and an antibiotic should be administered following the copper treatment. Copper levels may be changed by the pH of the system, staying in solution longer at a high pH. It is possible to maintain a safe copper level at a low pH and increase the copper to toxic levels by doing a water change that significantly raises the pH of the water and re-releases absorbed copper back into solution. Chelated copper will remain in solution longer than ordinary copper sulfate. Always use high quality copper to avoid contamination from other minerals that may prove to be toxic.

Tetra medicated food, flake "D", is highly recommended in conjunction with copper, and in a system with invertebrates may be the only available treatment. (Copper kills invertebrates).

Signs of a Parasite Infestation are:

- Visible spots, usually white, that make the fish look like has been salted or covered with powdered sugar.
- Rapid or heavy breathing. Some parasites will attack the gills before any can be seen on the fins or body, and the fish may die from suffocation.
- Scratching. If a fish constantly rubs against objects in the system and looks like he is trying to dislodge something, he is probably trying to rub something off and it is probably parasitic.

Because ICH reproduces in the system, the whole system must be treated, not just the infected fish. Hospital tanks and dips seldom effect a permanent cure, and cause a great deal of extra stress for the fish. Ich infections may produce an immunity to later attacks.

Malachite Green is effective against parasites but may damage gills and burn the fish. Malachite green that is not pure is especially dangerous.

Dylox (DTHP) is effective; it is an insecticide and is generally not toxic when used as directed.

Oodinium: This is actually a form of algae parasitic on fishes. Treatment is described under ICH.

Anchorworms: These are easily visible and look like little sticks about 1/4" long protruding from the body or fins. They are firmly attached and when pulled out may hold onto a piece of flesh. Medicated food, flake "D" and Dylox is the recommended treatment and it usually takes 10 to 14 days for full eradication. After a few days of treatment, any remaining worms should be removed from the fish. Because of the large sores left by the parasite, fresh water and an antibiotic is a must during and following treatment with Dylox and flake "D".

Fish Lice: Are crustacean parasites with similar treatment as per anchor worms. They are about one quarter inch flattened discs with rasping mouth parts and hook-armored legs capable of damaging fins and skin. Treat for secondary infections as well as eradicating these pests.

Lymphocystis: This is a virus that lives off of impurities in the water while attached to a fish. It does not live off the fish (like ICH), but may kill indirectly by interfering with gill movement, swimming ability, or eating. Lymphocystis can only be killed in an established system by removing its food source by means of purifying the water. This can be done with ultra-violet sterilizers, ozone, chlorine, frequent and large water changes, micron filtration, or diatom filtration. In the late stages of successful treatment, the virus clumps may be easily removed from the fish by scraping with your fingers, or may drop off as it dies.

The virus itself cannot be trapped by filtration, only its food source can be removed. Lymphocystis virus transfers easily between species but less easily between genera. The more crowded an system, the easier the virus may transfer from fish to fish.

Bacteria

Bacteria grow erratically and are often white or milky in appearance. A bacteria infection may be localized or may be evident on several areas of the fish. Bacteria infections are likely to be found in or around open sores or any area where the fish has lost it's protective slime coating. Antibiotics and medicated food should be used to treat bacteria infections along with frequent water changes. A dirty system can prevent successful treatment. Because there are so many different types of bacteria, you may have to try several types of antibiotics before finding one that works. Be sure to do large water changes between treatments of different medications. Different resins and high-quality carbons will remove medications from the water and should be removed during any treatment with antibiotics. Internal bacterial infections may be identified by the gas they produce. This may cause the following symptoms- swelling, a fish that has trouble staying on the bottom, whitish feces that float or trail off behind the fish, or lack of feces entirely (blockage).

Fungus

Fungus spreads evenly, starting from a central point and growing in an outward pattern. Several areas may grow outward until they overlap and give the appearance of a bacteria infection. Fungus is white with a velvety or even hairy appearance. It is most likely to be found on the mouth, eyes, or tips of the fins. Treatment consists of water changes, medicated food, and antibiotics. Sulfa drugs administered through food may be your best treatment.

Popeye

This is a symptom, not a disease caused by a specific organism. It is manifested by swelling behind the eye(s), or in the eye(s). The swelling may be caused by many factors but is most commonly caused by bacteria. If unilateral (one-sided), the cause is probably mechanical injury. Only time passing may effect a cure. It is difficult to treat, but the most effective procedure seems to be a good environment, and medicated food. Start with flakes "D" and "A" in conjunction, then feed flake "B", and go on to flake "C" if needed. Erythromycin or Chloramphenicol may also be effective.

Swim Bladder Disease

The swim bladder is the organ which allows a fish to stay at any level in the water column without sinking or floating. The swim bladder may fail from damage by bacteria, parasites, genetic faults, or blows and/or bruises. When the swim bladder fails to function the fish loses it's ability to swim normally and may swim sideways or even upside down. Once damaged, the bladder does not usually return to normal function, but if the fish can eat and swim without too much strain it can live for years with the condition. Goldfish seem particularly prone to swim bladder problems. Without knowing the exact cause of the malfunction, treatment is difficult. Since internal bacteria, fungus, or parasites are the only treatable causes, medicated food and/or antibiotics should be tried along with frequent water changes.

Dropsy

Dropsy is a name given to any disease that causes a fish to swell so much that the scales no longer lay flat against the body of the fish. By looking down on a fish you can easily spot a case of dropsy. This is a very difficult condition to treat successfully. Daily doses of Erythromycin, daily water changes, and exclusive feeding of medicated food (D&A) has proven to be the most effective treatment. Including regular feedings of daphnia may prevent some cases of dropsy. Goldfish have the most trouble with dropsy but it may be found in any fish.

Suffocation

Rapid breathing or gulping near the top of the tank may mean a fish is not getting enough oxygen. This may be caused by:

- No air circulation. Air pump or stones may be faulty or missing. Surface agitation may be missing in a "system".

- Temperature is too high. The warmer the water, the less oxygen it can absorb and hold.
- The water surface is covered. Water cannot absorb oxygen if the surface is covered with scum or oil.
- Parasites. Gills clogged with parasites cannot get oxygen. Even after parasites are killed, if they are so numerous that they remain clogged in the gills fish will still suffocate.
- Overmedication burning gills, rupturing blood cells, causing too much mucus production.

Brain Damage

Fish may show any unusual symptoms. This should be only offered as a diagnosis after all other possibilities have been ruled out. Erratic, jerky swimming or spinning are common signs of brain damage. Brain damage can be caused by parasites, bruising (concussion), high or low temperatures, or toxins.

Toxins

Symptoms look the same as brain damage, but all or most of the fish in the system are affected at once. Spinning is the most frequent sign of a toxin. Common toxins are : Windex, ammonia, paint fumes, bug sprays, copper, flea and tick collars, (by hand contact), colognes or perfumes, and chlorine. Water changes is the most effective methods of removing poisons quickly.

Open Sores

These can be caused by:

- PH that is too high or low.
- Copper burns.
- Shipping damage.
- Scraping on rocks or other objects.
- Bites.
- Parasites.
- Internal infections reaching the outside.
- Net damage during handling.

Treatment consists of eliminating the source of the problem, and the use of Novaqua, Polyaqua, or any other artificial coating medication. Be sure to watch for signs of bacteria or fungus and treat accordingly if it appears. Salt in the water can be a very effective prevention of fungus or bacteria infection. Clean water will also retard the growth of infection.

NOTE: Many fish diseases are not treatable - either is no known cure or the cure may jeopardize the fish. Some fish die from internal parasites that are undiagnosed and therefore untreated. Do your best to help your system and it's fish, but remember that we cannot do ever...

Treatment

Oxytetracycline

Dosage: 2 teaspoons (7.5 Grams) of Oxytetracycline Powder for every 1 pound of koi pellets. Mix the medication with approximately 2 ounces of water. Add this to a spray bottle, and spray the entire contents over the feed. Let air dry. Note: It is important that you do not use any heat element to dry the feed because it will destroy the medication. Use a fan to expedite the drying process if necessary. Feed the medicated feed to the fish once a day for two weeks. Do not feed any other food except for the medicated food during treatment time. You do not need oil or gelatin to get the medication to stick to the food. This is a waste of time, and it pollutes the water.

Forma-green

We use this solution at 2 drops per gallon, which works out to be 100ML per 1,000 gallons (milliliter measurements can be found on a standard kitchen measuring cup). Dosing this product may vary depending on your water composition. In some areas the water will stay green colored for a week, and in other area's the green is gone in 24 hours. It is important to keep the water green colored, during the entire duration that you are using the antibiotic (which will be around 2 weeks). Plan on using the Forma-Green once every three days. This will prevent a secondary infection from developing in the sores.

Also, remember to sterilize all of your pond equipment with bleach to prevent spreading the disease to any other tanks or ponds. It is also important to obtain a gram scale (diet scale) which can be found at most local retail stores for about $7.00. These medications need to be accurately measured out for effectiveness and to prevent over treating the pond.

Sometimes cases of Aeromonas are so severe that no treatment can reverse the damage done by the bacteria. In these cases I suggest that you remove the infected specimens and give them a humane solution to their suffering by putting them in a bag of water, and then put them in the freezer. We have had customers call us and say that they see bones sticking out of the fish. This is an example of a fish that cannot be saved.

Important Differences in Bacterial Infections with Koi and Goldfish

Ok, so you saw that picture above with the large ulcer on the Koi. There is another bacteria that will do just as much, or more damage to your fish that is called Pseudomonas bacteria.

The way to tell the difference is, that with the pseudomonas bacteria: You will also have fin and tail damage. To treat for pseudomonas, instead of using the treatment listed above... use Neomycin or Kanamycin in the feed instead. Also use the Forma-Green in the water.

Branchiomycosis

Infected fish show signs of asphyxia such as gasping. They appear weak and lethargic. Die-offs can reach a mortality rate of over 50% in some cases. The fish that survive, may be able to re-generate their damaged gill tissue if treated properly.

This disease is seen in the blood vessels of the gill tissue, and it obstructs the circulation of blood through the gills, which makes the gills lose their bright red color. The gill tissue becomes mottled with patches of brownish discoloration due to hemorrhages mixed with whitish areas of disrupted blood flow, resulting in necrotic tissue. In some cases, the necrotic areas may slough off and allow the gills to be invaded by a secondary disorder such as a Saproglenia fungus.

Gill infection with Saprolegnia fungus looks almost identical to Branchiomyces. This suggests that Branchiomycosis may belong to the family "Saprolegniaceae".

Mycotic infections in goldfish and koi are mainly associated with stress or injury. Infection with Branchiomyces is rapid, and is mainly caused by algal blooms, overcrowding, high water temperatures, and high levels of ammonia in the pond.

Prevention of Branchiomycosis can be accomplished by maintaining good water quality, removing dead fish and preventing the accumulation of decomposing organic matter. Decreasing feed, and improving water flow to prevent the build up of ammonia are also crucial. Thinning out fish to prevent crowding and stress should also be practiced.

Treatment and Control: A long term bath in Acriflavine Neutral or Forma-Green for seven days helps this condition. If you decide to use the Acriflavine Neutral, you might want to do this in a holding or quarantine tank because it's awfully hard to remove from the water.

Water changes and the reduction of algae with the use of a nitrifying facultative bacteria like Aqua Gold can prevent Branchiomycosis and many other diseases like Aeromonas (hole in the side disease) from developing in koi ponds.

Remember: Water quality is key, in any environment that contains tropical fish.

Columnaris Disease

In many cases, we receive calls from customers stating that they have white wavy worm-like parasites attached to the glass in the aquarium that tend to sway back and forth with the water circulation in the aquarium. Often, the fish do not seem to be affected at this stage. Given time for this disease to spread, the infestations usually begin on the fins, which usually become frayed and ragged. The disease will spread to the skin, eventually causing ulcerations and irregular areas of epidermal loss. Aeromonas hydrophila is commonly present in advanced lesions and contributes to the pathology.

In the gills, Flexibacter Columnaris will color them light to dark brown and you will also notice some necrosis. On the skin, the fish will appear to have mold growing on it, with a slight cottony look, due to a fungal infection that has attacked the lesions and ulcerations. The lesions and ulcerations in advanced stages are usually infected with a secondary motile aeromonad. So as you can see, here is a situation where you have multiple infections present.

Flexibacter Columnaris can persist in water for up to 32 days when the hardness is 50ppm or more, but a hardness of 10ppm reduces viability considerably. The addition of carbon to the system increases the survival of this disease in hard water, but this is not the case in soft water.

Columnaris is prevelant in systems with high organic loads, crowded conditions, handling and low dissolved oxygen content. Lesions generally develop in 24 to 48 hours following handling, followed by death at 48 to 72 hours if not treated.

Treatment and Control: Any Sulfa drug combination will work well. TMP Sulfa, Sulfa 4 TMP, or Triple Sulfa.

As with all diseases of tropical fish, proper maintenance and water quality are the key to success.

Heteropolaria

Here is a very interesting article for all Koi hobbyists and professionals. Everyone who has ever dealt with Aeromonas (hole in the side disease) should read this article.

Heteropolaria affects the body, fins and gills of the fish. This disease is often associated with Aeromonas and Pseudomonas. Epistylis can be the cause for these two, and many other diseases.

This disease is common in waters containing high organic matter, i.e. ponds and lakes. It is prevalent during winter months, but outbreaks of red sore disease are more commonly seen in spring and summer months.

Heavy infections of Heteropolaria are characterized by white-gray cottonlike patches on the body surface. Ulcers may develop, and the fish may roll on their sides and start "flashing" like they have parasites. Red sore disease is also characterized by scale erosion, lesions on the body surface and bacterial hemorrhagic septicemia. The infectious diseases are actually caused by the attachment of this protozoan.

So basically, this protozoan may be the cause of secondary infections in your fish and must be dealt with accordingly. Water quality is always the key, and infections like this may be prevented by maintaining good water quality. Heteropolaria may be treated by using Formaldehyde or Forma-Green in the pond. Metronidazole or Quinine Sulfate may also be used to combat this protozoan infection if you isolate the fish and treat him in a hospital tank.

If you use a nitrifying bacteria, like our Aqua Gold to maintain your water quality, heteropolaria and many other diseases may be avoided. Earlier in this article we discussed the fact that heteropolaria is present in waters containing high amounts of organic matter.

Koi ponds contain enough organic matter to create a bacterial, fungal and protozoan "Factory" in most cases when not properly maintained. This is why many people are "constantly" treating their pond month after month and never have any success with their fish. The dirtier the pond, the larger breeding ground for disease is developed. We know that summer months are especially difficult due to algal blooms and soaring temperatures. Try to have some shade for your pond by planting some non-disciduous trees around it, or using a cover (many types of covers available through your local nursery).

Make sure you use plenty, even double or triple the amount of nitrifying bacteria in your pond during the summer months. Also use an algal inhibitor like Crystal Blue if you have problem green water. Also make sure your filtration is in accordance with the load of fish that you have in the pond.

Overfeeding is another cause of undue organic matter. You don't need to feed your fish 3 times a day. Once a day is just fine. Feed enough food so that they can clean it up in about 5 minutes. More than that is considered over-feeding.

Now, most importantly, we discussed the fact that heteropolaria can cause an opening for a secondary bacterial or fungal infection to happen right? O.K. so you have ulcerations on the fish. I suggest first that you try something like Oxytetracycline Hydrochloride in the feed. Use 2 teaspoons mixed in approximately 2oz. of water. Mix well. Add the medication to a spray bottle and spray it over 1 pound of koi pellets.

Let air dry (not in the sun, oven or with hair dryer), use a fan to expedite the drying process if necessary. Once the food is completely dry, feed this to the fish once a day, for up to 14 days. This treatment is for hemorrhagic septicemia which Koi are very susceptible to.

If the Oxytetracycline produces no results, you may have Pseudomonas. If this is the case, use Neomycin or Kanamycin Sulfate Powder at the same directions and treatment method explained above. With both diseases, isolate the fish if possible and use Forma-Green in the water to help prevent a secondary fungal infection like Saproglenia from developing.

A temperature bump also speeds up the healing process. Bump it up to 75-80°F if possible. We have cured fish with these diseases in two weeks using these methods.

Hexamita: Internal and External

The external form of Hexamita is referred to as: Hole in the head disease. This disease affects susceptible fish such as discus, angelfish and oscars, which seem to be the most popular carriers of this problem (note that this disease may affect all fish).

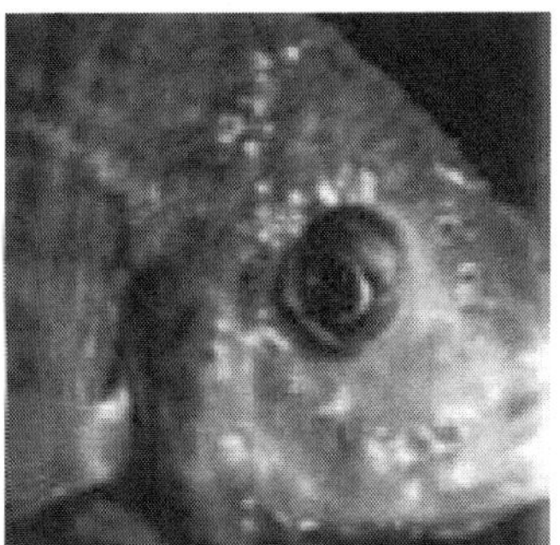

Figure: *External Hexamita*

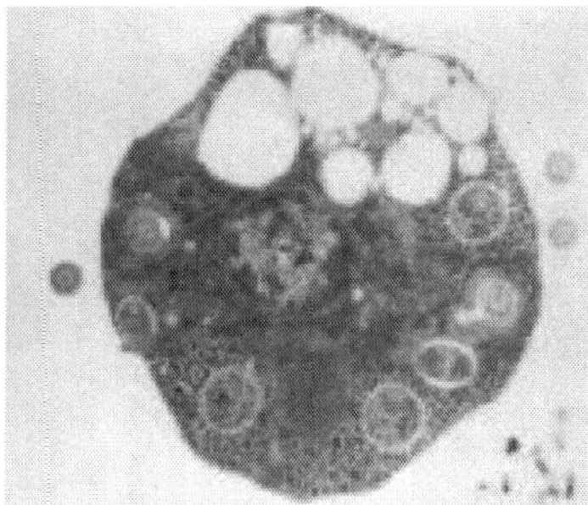

Figure: *Internal Hexamita*

This disease usually starts out as a small "pimple" on the head, and as the condition proceeds in severity, ends up to be a very large sore causing lesions in the epithelium and... eventually ending up in death of the fish. Sometimes you are able to see small white nodules, sticking up out of the sore. The lateral line is also another area where this parasitic protozoan can be seen. This disease is best treated early on with Metronidazole, or something stronger like Quinine Sulfate.

The internal form of Hexamita are flagellated protozoans found in the gastrointestinal tract of a wide variety of fishes. They frequently infect discus. Spironucleus may be a distinct organism from Hexamita, as it is longer and possibly more sinuous, but for practical purposes, both organisms appear to cause similar clinical responses. These parasitic protozoa are very motile. The flagella are usually not easily seen. Many times infections are not apparent. In angelfish, discus and gouramis, the disease is characterized by poor condition, weight loss and death. The fish may also show excessive nervousness, turn dark in color, and hide in the aquarium.

Again, Metronidazole is the drug of choice for internal Hexamita. Use 1 teaspoon per pound of food (frozen food is preferred). Thaw the food and mix the Metronidazole into it. Return the food to the freezer, and once frozen you are ready to start the treatment. Feed it to the fish once a day for a minimum of ten days. Do not feed the fish any other foods during this treatment. The treatment may take longer according to the condition of the fish, and the severity of the disease.

Metronidazole is for internal Hexamita, and the Quinine Sulfate is for external Hexamita only.

Mycobacteriosis

Mycobacteriosis is worldwide in distribution. All fish species should be considered susceptible. Some are more susceptible than others, like gouramis, neon tetras, discus and labyrinth air breathers.

Clinical Signs: Mycobacteriosis is a chronic progressive disease. It may take years for it to develop into a clinically apparent illness. Some signs to look out for include: Lethargy, anorexia, fin and scale loss, exopthalmia, emaciation, skin inflamation and ulceration, edema, peritonitis and nodules in muscles that may deform the fish.

Examinations usually reveal gray or white nodules in the liver, kidney, heart or spleen. There also may be skeletal deformities. Diagnosis is usually based on clinical signs and the presence of acid fast bacteria in tissue sections.

Mycobacteria are gram-positive, pleomorphic rods that are acid-fast and nonmotile. They form cream-colored to yellow colonies on solid media. It is suggested that transmission of this disease may be caused by contaminated food infection rates can be quite high in contaminated freshwater tropical fish production facilities.

The aquatic environment is considered the reservoir. Mycobacterium marinum has been cultured from swimming pools, beaches, natural streams, estuaries, tropical fish tanks and city tap water. Human epidemics of granulomatous skin disease have occurred from swimming in infected water. This mode of human infection is much more common than infection from exposure to infected tropical fish tanks. This disease will usually attack a sore or abrasion and be apparent about 2-3 weeks after exposure.

Treatment and Control: Kanamycin + Vitamin B-6 for 30 days is the most effective treatment that we know of for tuberculosis. The fish should be quarantined during treatment time. Liquid baby vitamins found at your local pharmacy are a good source of vitamin B-6. One drop per every 5 gallons of aquarium water is sufficient.

Replace the vitamins according to how much water is changed in the tank during treatment time.

Overcrowding and poor water quality are usually the cause of this disease.

Be careful, this disease may spread to humans.

Water Mold Infection

Synonyms. Saprolegniosis, Ulcerative mycosis, Oomycetes, Cotton mouth disease.

The most common presentation of water mold infection as as a relatively superficial, cottony growth on the skin or gills. Such lesions usually begin as small, focal infections that can rapidly spread over the surface of the body. It is not unusual for large lesions to suddenly appear within 24 hours. New lesions are white, due to the mycella of the fungus, and over time will become red, brown, or green as a result of trapping algae or debris. When the fish are removed from the water they appear to have a “slimy” matted mass growing out of the skin and scales.

This disease will progress rapidly, producing lesions that can form a deep, necrotic ulcer. This can extend deep into the body and will frequently affect internal organs. After time the infected tissue will slough off, leaving a large crater-like hole, surrounded by dark red or white muscle.

Although typical saprolegniosis lesions grow rapidly over the surface of the skin, they usually do not penetrate deeply into muscle. However, the damage to the skin or gills may be enough to kill the fish. The severity of the disease is determined by the area of skin and gill damage. The larger the area affected, the greater the osmotic stress and electrolyte imbalance. Skin woulds due to net damage or other trauma increase the risk of infection.

Many oomycetes display a seasonal re-occurrance. Saprolegnia species are seen mostly in cooler months of the year. Most saprolegniaceous oomycetes are prevented even by moderate salt content in the water. There is also evidence that many fungal infections affect hosts that are “stressed”.

Oomycetes infections can also be a secondary infection. If the fish had a pre-existing bacterial infection due to a motile aeromonad (aeromonas or hole in the side disease), the fungal disease will surely attack the open sores.

So, basically... there are many factors which can contribute to this disease. It is suggested to research thoroughly all contributing factors before diagnosis is made.

Treatment and Control: Forma Green is good at controlling algae, molds and fungus, but won't always work on this problem (depends on the severity). Erythromycin would be the best for treating this problem. Erythromycin will also kill cyanobacteria in marine tanks (red algae), and black algae Aspergillis Niger.

Koi Pond Medicating Techniques

Hospital Tanks

Any new fish to be added to the pond should be observed for at least two full weeks. Always assume a new fish is diseased with bacteria or parasites. Be especially careful if adding "feeder goldfish" to the pond. Since these fish are bred to be part of the food chain, less care may have been taken to ensure their good health. If the new fish are the only fish in the pond, the entire pond may be treated. Try to avoid treating the main pond as it contains crucial nitrifying bacteria, which may be destroyed by medications. It is always more manageable to treat fish in a smaller, more observable body of water such as a glass aquarium.

A glass aquarium may be set up as a hospital tank. It should be equipped with an aeration device, usually an air pump with an air-stone attached. A re-circulating filter may be used, but make sure that it only contains filter floss or foam... not activated charcoal. Activated charcoal will remove many medications that you may need to treat your fish.

To determine the fish holding capacity of the hospital tank, compute the square surface area of the tank in inches. Divide this figure by 30 to determine the number of body inches the tank can accommodate. Since overcrowding is also stressful, it is important not to add this factor to an already stressed and sick fish.

Using Med's In The Hospital Tank

New fish should be treated with a broad-spectrum anti-bacterial, anti-parasitic medication as a preventative. Antibiotics should never be used unless a specific disease is noticed. Some preventative medications include: Acriflavine Neutral, Copper Sulfate, Forma-Green, Malachite Green, Methylene Blue, Praziquantel and De-Los. One or two tablespoons of salt per 5 gallons will aid in relieving stress as well as in treating for some parasites.

The water temperature in a hospital tank should be maintained at a stable level close to that of the pond. However, if a salt only treatment is being used to destroy possible parasites, you may want to use an aquarium heater and heat the tank up to 86°F.

The temperature must not be raised more than 1 degree per hour to avoid stressing or damaging the fish. This high temperature will kill most parasites, and also kill S.V.C and K.H.V. viruses.

Make sure that you allow the water to cool down slowly, back to the temperature of the outdoor pond before you transfer fish into it. Also make sure to sterilize all equipment and nets with a watered down bleach solution, that were used to transfer the fish into and out of the hospital tank. Cross contamination can be a significant problem.

Temporary Fish Quarters

In treating a large fish or a number of fish, the use of an aquarium may not be practical. Temporary quarters for fish may be built of wooden planks supported by concrete blocks to accommodate a water depth of 9 to 12 inches will facilitate volume computation for medication dosages. A double sheet of polyethylene is draped and secured in the frame. Shade should be provided in warm, sunny weather. A net or a screen over the top will prevent fish from jumping out and predators from getting in. A small pool pump or aquarium air pump helps to maintain an oxygen supply.

Using Antibiotic Treatments

First of all, you never, ever want to use an antibiotic treatment for a preventative. You can create problems, and actually make the fish antibiotic resistant by doing this. Also, never use an antibiotic for less than 10 days. If you do, you may create a resistant strain of bacteria to develop. Antibiotics should never be used as "dips" either. These drugs must be used as a long-term bath for 10 days to be effective. Antibiotics may have to be used longer than this, especially if you are treating an outdoor pond that contains cooler water. It is always better to remove the fish to a heated hospital tank or pool for treatment. This will speed up healing time.

Adding Salt to Your Pond

Salt is pretty amazing in it's ability to control algae, detoxify Nitrites, kill parasites and it's antiseptic qualities. Salt is a great item to use for your water quality, but first... you need to know how much to add. We feel that a 0.1% continual salt bath is a good level to run at all the time. To achieve this level, add 1¼ ounces of salt per 10 gallons of pond water.

The maximum level of salt that you can run without major damage to the fish is 0.3%. This high salt level is used for treating fish wounds and parasites. To achieve this level, add 3.8 oz. of salt per 10 gallons. This salt level is better suited for a bath, or in a hospital tank.

Never ever take your main pond up this high as a long term bath in a high salt concentration is very bad for your fish, not to mention your biological bacteria. You will ruin your pond and slowly poison your fish with all of this salt. Trust us, around 50% of the calls we are getting are salt-related problems.

Name: DROPSY Symptoms:

- Big fat belly, not pregnant
- Huge Swelling of Body
- Bulging sides and stomach
- Scales almost popping off
- Scales may be forced outward
- Eyes may pop out
- Symptoms of Stress & Disease Dropsy General Description Dropsy is not really a disease. For all intensive purposes we will call it one. It's really an internal bacterial infection usually caused by poor water quality.

Fish may recover with no treatment and may die despite it. There are multiple possible causes. Sometimes it's not contageous, but sick fish should be isolated and treated since determining the actual cause may be impossible.

The swelling is because the fish is absorbing water faster than it can eliminate it, and it can be caused by many different problems. High nitrates are one thing to check. Internal bacterial infections, including fish TB, are other possibilities. If there are no water quality problems, you may want to attempt antibiotic treatment in a separate tank.

There are multiple possible causes. Usually caused by kidney damage. Kidney damage may be caused by overuse of drugs or a disease. Eventually the swelling will cause the scales to raise, giving the fish what is called the "pine-cone" appearance.

You can best see this by viewing your fish from the top. Fish may also stop feeding, appear off-colour, become listless and/or lethargic, have sunken eyes, and hang at the top or stay at the bottom of the aquarium. Dropsy Treatments Dropsy is not very contagious; however, Fish usually die from this, but in some cases where the problem is due to bacteria, if detected early enough, it can be treated.

It's possibly the hardest internal bacterial infection to cure. There are a number of medications available such as penicillin, tetracycline and naladixic acid. The fish usually doesn't make it. By the time the scales begin to raise, however, it is very fatal to the fish. Salt baths can help to draw the fluid out of the fish. A variety of medications can be purchased that treat dropsy, which sometimes occurs due to an internal bacterial problem. Medications for external bacterial problems only will not be effective for this problem.

Dropsy Prevention Poor water conditions are often the culprit. Gouramies, Cyprinids (barbs, danios, etc), guppies, betta and goldfish are prone to this disease. Goldfish are said to be somewhat more prone to dropsy than other fish. High nitrates are usually the culprit. Clean Water, is a must! Clean Water, should I say that again? Good water conditions prevent this.

Aquarium Fish Health: White Spot Disease Symptoms and Cures

Fish death is one of the main problems that beginner aquarist and even some expert aquarist face. Itÿÿs frustrating to the extent that most quit keeping aquarium fish.

But fish death can be avoided. Most fish deaths are caused as a result of both an internal and external types parasites that compete with the fish in tank.

As a result if you watch your aquarium fish often you should be able to discover when they have been infected by this parasite and be able to treat them to avoid fish death.

Look out for the following White Spot disease behavioral symptoms in your fish.

- Constant lying on the bottom or hanging at the surface.
- Rubbing of the body against rocks
- Gasping at the water surface
- No response to feeding
- General dullness and lethargy

- Hovering in a corner
- Fish swimming with clamps up

The most common of the visible signs is the development of the pin head-size while spots on the body or fins. This ailment is referred to as White Spot disease and is caused by the parasite - Ichthyophthirius Multifillis.

This parasite has a free-swimming stage, which attaches itself to the fish. The most common chemical used in treating infected fishes is Methylene Blue. You could buy a one per cent stock solution from a reputable chemist or aquarium shop and apply at 0.8 to 1.0ml per gallon of water. This amount should be added all at once. Repeat after one or two days.

The fishes must remain in this bath until every while spot has disappeared. A water change after treatment is necessary or else prolonged contact with the chemical may affect the fertility of the fish.

Another tip if you are using a side filter with activated charcoal should remove it to prevent the coal from absorbing the Methylene Blue.

Another tip... during treatment you should use artificial aeration with coarse bubbles near the surface, since a dirty bottom would inactivate the medicament by absorption. A better measure is to remove all dirt from the bottom before treatment.

Methylene Blue is harmless to young fish and unlike the general belief, it does not affect plants if used in weaker concentration.

Aquarium Fish Health: Dealing With Cotton Mouth Disease (Mouth Fungus)

Cotton Mouth disease also know as Mouth Fungus is a disease your fish can get and it needs to be dealt with quickly. Cotton Mouth disease is not as common as the while spot disease, but, it is highly infectious and contagious.

The victim fish shows a whitish fungus round the cheeks and lips. The lips may become swollen and rot away. Sometimes a rotten strip of lip attached only at one end will move in and out of the mouth as the fish breathes.

Fish infected with Mouth Fungus lose their appetite and their movement become sluggish. If no adequate treatment is given, the whole frontal part of the head may be eaten away finally and the fish dies.

Unless the affected fish is of consideration value, it should be killed before this fatal disease attack sthe other occupants, of the tank. Think about it... is trying to save the life of one fish worth risking the death of the rest of the fish in your aquarium?

But if you insist on keeping the fish or in case the infection has already been passed on to other occupants, the following treatment is advised:

- Swabbing the mouth of the victim fish with a soft cloth dipped in strong salt solution. Then you must then keep the patreat isolated in a bucket or jar containing a strong salt water.
- Try swabbing the lips with a 5 per cent silver mercury preparation.
- Make a solution of Terramycin or Aureomycuin by dissolving 50mg per gallon of water, a rapid cure is expected within 48 hours.

You can try all of the above remedies, but the most common remedy is the popular Methylene blue solution. To perform this remedy the sick fish should be placed in a jar, bucket or a treatment tank into which has been added a methylene per blue to colour the water deep blue.

How to Discover and Prevent Aquarium Fish Illness

Aquarium fish fall ill just like any other pet. The illnesses are as a result of disease. The common diseases that affect aquarium fish are mostly stress induced.

The micro-organisms which cause these diseases may be present in the water as part of the normal micro fauna and grab the opportunity to infest or infect the fish when it is stressed and its normal defense are weakened.

Different Ways Aquarium Fish Suffer from Stress

- Aquarium fish usually start experiencing stress starting from the time itÿÿs about living the breeding farm till when the end purchaser finally picked it from retailer. Most beginners donÿÿt really know how to handle fish.
- The quality of water is another cause of aquarium fish stress. Different fish species have different water quality that will make the environment conducive for them. Quality parameters like pH, water hardness, high nitrite and carbon dioxide level,

low dissolved oxygen salinity of water, water temperature and others.

- Any change in environment like physical damage, leaches, lice and introduction of new fish that is not quarantine could lead to stress in aquarium, thus making life non conducive for fish in aquarium tank
- Change in weather condition is another factor that leads to aquarium fish stress. When there is heavy downfall as you know the weather will become cold and this could lead to stress in aquarium.
- Poor Diet as a result of lack in nutritional requirement always leads to poor performance in fish immune system. Fish immune system will struggle to operate efficiently. This will result in a stressed fish.

Warning Signs of Illness in Aquarium Fish

You can easily know when your fish are going through stress if you watch them very well and often. You will notice signs like fish swimming with clamps up (closed) in their fins, hovering in a corner, heavy breathing and fish brushing its body against objects. These are warning signs you need to act on immediately.

How to Reduce Stress in Aquarium

To reduce stress in aquarium, you need to guide against factors that lead to stress. The following are tips on how to prevent stress that ultimately leads to illness:

- Make sure you have detail information about the fish species that you will be introducing to your tank. Enquiring about its requirements before you buy it will help you a lot. Thus, ask lots of questions before making your purchase.
- Knowing the right food for your fish is another important thing you have to know as this may vary from species to species. If you have this knowledge you will be able to provide foods that are rich in nutritional ingredients that your fish require.
- Keep the environment clean.
- When you want to introduce new fish, make sure it's quarantined to reduce the risk of disease.

Toxoplasmosis and Associated Diseases and Disorders

Toxoplasmosis is considered to be the third leading cause of death attributed to foodborne illness in the United States. More that 60 million men, women, and children in the U.S. carry the Toxoplasma parasite, but very few have symptoms because the immune system usually keeps the parasite from causing illness. However, women newly infected with Toxoplasma during pregnancy and anyone with a compromised immune system should be aware that toxoplasmosis can have severe consequences for them.

The parasitic zoonosis toxoplasmosis, which was poorly understood before the advent of the HIV epidemic, has become a major clinical problem worldwide. Toxoplasmosis is caused by a coccidian parasite Toxoplasma gondii. Humans acquire toxoplasmosis from cats, from consuming raw or undercooked meat, water, transfusion of infected blood, organ transplantation, and from vertical transmission to the fetus through the placenta during pregnancy. Studies of the unique environmental factors in various communities indicate the important roles that eating habits and culture have on the transmission of this infection.. Cockroaches may excrete *Toxoplasma* in their feces for up to 10 days after ingesting infected cat feces.

Recent advances in understanding toxoplasmosis have been made in the areas of the basic biology of the parasite and the host-parasite interaction, especially the cellular immune response. There is new insight into the biology of the cyst stage that is responsible for meat-associated transmission of infection and for the reactivation of disease in chronically infected humans. Fewer recent advances have been made in clinical diagnosis and treatment of toxoplasmosis. The fascinating revelation that Toxoplasma gondii contains an organelle, now known as the apicoplast, that derives from an algal endosymbiont, has opened many avenues of basic investigation. An understanding of the fundamental biology of T. gondii promises future progress in prevention or treatment of toxoplasmosis.

Cases of 2 consecutive siblings with bilateral macular lesions, for which there is strong clinical and laboratory evidence supporting the diagnosis of congenital ocular toxoplasmosis. These cases raise the possibility of maternal parasitemia during Toxoplasma gondii reinfection, leading to transmission to the fetus and congenital ocular toxoplasmosis despite prior immunity and lack of an immune disturbance in the mother. A 38-year-old woman who had been treated for ocular toxoplasmosis 20 years earlier delivered a newborn who presented with a focal necrotizing retinochoroiditis characteristic of toxoplasmosis, as well as positive immunoglobulin (Ig) G and M serology for toxoplasmosis.

The workup was negative for other entities. CONCLUSION: This case suggests that women with old retinal scars due to toxoplasmosis and long-standing IgG antibodies to toxoplasmosis are also at risk of transmitting this disease to the fetus.

The development of glomerulonephritis as a complication of neonatal lupus or congenital toxoplasmosis is extremely rare. We report a case of membranous glomerulonephritis occurring in a neonate with toxoplasmosis and atypical congenital lupus, including high-titer antinuclear antibodies and hypocomplementemia. This case illustrates that glomerulonephritis in the neonate may be induced by passively transferred maternal antibodies.

Congenital Toxoplasmosis

Congenital toxoplasmosis is a potentially serious infection which usually affects infants born to non immune women. It may lead to severe visual impairment or neurological complications in the child. Neonatal screening is of importance to diagnose children with no clinicl signs as treatment has been shown to reduce long-term complications. Ophthalmological investigations should start early and continue in co-operation with paediatricians. A baby with congenital toxoplasmosismay be born to a normally immunocompetent woman previously immunized against toxoplasmosis. Therefore, toxoplasmosis cannot be excluded on the ground of maternal immunity status and must be quickly investigated, given the emergency of appropriate treatment.

Congenital toxoplasmosis secondary to maternal primary infection acquired late during pregnancy is generally can be easily diagnosed at birth. A newborn infant was born to a woman who had been infected between the 27th and the 33rd week of gestation. No treatment had been given during gestation. The infant had a disseminated form of toxoplasmosis with hepatosplenomegaly, pneumonitis, purpura, hepatitis. On the third day of life, he developed shock. The patient died early despite therapy. Septic shock is unusual in congenital toxoplasmosis, although it has been described in immunocompromised patients, notably in patients infected with the human immunodeficiency virus.

Reactivation of Ocular Toxoplasmosis After LASIK

A 34-year-old man who underwent bilateral LASIK developed a toxoplasmosis scar in the retinal periphery of the right eye. After an operation, his vision improved, however, 52 days after the procedure, he complained of loss of visual acuity in his right eye. Examination revealed signs of anterior uveitis, vitreitis, and active chorioretinal lesion satellite of the old toxoplasmosis scar. The patient was treated with a multidrug regiment.

Toxoplasmosis and Retinochoroiditis

Toxoplasmosis is the most common cause of infectious retinochoroiditis in otherwise healthy individuals. Most cases of Toxoplasma infection in the immunocompetent adult have no clinical signs. The most common clinical presentation is localized enlarged lymph nodes.

Ocular signs, which are common in congenitally acquired toxoplasmosis, may rarely be the only manifestation of acquired systemic toxoplasmosis. It has been suggested that concomitant infection with a DNA virus, such as CMV or herpes simplex virus, may facilitate the penetration of protozoa into cells, or that antigenic stimulation from toxoplasma antigens may activate hidden CMV in the affected person.

In case of systemic toxoplasmosis, patients usually present with enlarged lymph nodes, general malaise and rash which progress to ocular signs and symptoms of retino choroiditis. Affected individuals are treated with several medications (pyrimethanime, sulfadiazone and folic acid, plus prednisone) for several weeks, during which the illness usually resolves.

Toxoplasmosis with Subsequent Hearing Loss

In immunocompetent patients the acquired toxoplasmosis is usually a mild or disease with no clinical signs. central nervous system manifestations are rare, most often in patients with HIV infection or in patients with other types of immunosuppression. However, it can also affect children and adults with normal immune system. In one case, a 9-year old healthy boy, who was hospitalized after one week with subfebrile temperatures and headache with clinical signs of encephalitis and unilateral deafness. He was diagnosed with toxoplasmosis and treated pyrimethamine and sulfadiazine. While most of the signs and symptoms disappeared rapidly the deafness persisted.

Toxoplasmosis Transmitted by Blood Transfusions

Blood from humans collected into heparin or citrate was inoculated with toxoplasma organisms. After storage at 4 C up to 28 days, samples were injected into the ear veins of rabbits. The test rabbits developed toxoplasmosis. Similar results were obtained by transfusing rabbits with blood obtained from rabbits subcutaneously injected with toxoplasma organisms.

Acquired Ocular Toxoplasmosis In Deer Hunters

Five young men presented with flu-like symptoms followed by visual loss due to a unilateral, focal necrotizing retinitis. All five men gave a history of ingesting undercooked or uncooked venison. All five had elevated toxoplasma, and all five improved clinically with an antitoxoplasma treatment.

In previously healthy young men, flu-like symptoms associated with visual loss and retinitis should prompt questioning about hunting and raw game meat ingestion, especially when toxoplasmosis is suspected.

Children with Malignancy and Toxoplasmosis

In one study aimed at the diagnosis of toxoplasmosis in 73 children with malignancy, 31 children had lymphoma (22 with Hodgkin's and 9 with non-Hodgkin's lymphoma) and 42 children had leukemia (34 with acute lymphoblastic leukemia and 8 with acute myelogenic leukemia). In positive cases toxoplasmosis was manifested by any of the following: fever, lymph node enlargement, neurological manifestations and/or enlargement of the liver and spleen.

The indirect hemagglutination test (IHA) for toxoplasmosis detected 4 (5.4%) positive cases with malignancy, 2 with Hodgkin's lymphoma, one with non-Hodgkin's lymphoma and one with acute lymphoblastic leukemia.

Toxoplasmosis and Cryptogenic Epilepsy. Recently, toxoplasmosis has been linked the cryptogenic epilepsies. Neuropathophysiology findings from various studies show a common physical relationship of microglial nodule formation in Toxoplasma gondii infection and epilepsy. This analysis raises the possibility that one of the many causes of epilepsy may be an infectious agent, or that cryptogenic epilepsy may be a consequence of latent toxoplasmosis infection. This raises the possibility that public health measures to reduce toxoplasmosis infection may also result in a reduction in epilepsy.

Toxoplasmosis and Risk of Schizophrenia

Because toxoplasmosis is known to adversely affect fetal brain development, some experts have suggested the relationship between maternal antibody to toxoplasmosis and the risk of schizophrenia and other schizophrenia spectrum disorders in the offspring. The findings may be explained by reactivated infection or an effect of the antibody on the developing fetus. Given that toxoplasmosis is a preventable infection, the findings, if replicated, may have implications for reducing the incidence of schizophrenia.

Toxoplasmosis Linked to Psychiatric Disorders

The activation of toxoplasmosis in individuals with AIDS has led to a realization the *Toxoplasma gondii* is among the most prevalent infection of the central nervous system throughout the world and has raised questions about what else it might be doing in human brains.

A national survey in the United States reported that 23% of people are infected. Recent research has focused on the possibility that *Toxoplasma gondii* may cause severe psychiatric disorders. Such research was stimulated by observations that AIDS patients who have reactivated toxoplasmosis have delusions, hallucinations, and other manifestations of disordered thought processes. There are also reports of such psychiatric symptoms in individuals with acquired toxoplasmosis; in a review of 114 such cases, "psychiatric disturbances were very frequent" in 24 of them.

Nineteen studies have been carried out that assess antibodies to *Toxoplasma gondii* in individuals with schizophrenia and other severe psychiatric disorders. All except one of the studies reported that patients had more antibodies than control groups had.

The parasite infects the brain by forming a cyst within its cells and produces an enzyme called tyrosine hydroxylase, which is needed to make dopamine. lead investigator Dr Glenn McConkey of the University's Faculty of Biological Sciences, believes that the findings could ultimately shed new light on treating human neurological disorders that are dopamine-related such as schizophrenia, attention deficit hyperactivity disorder, and Parkinson's disease.

Toxoplasmosis and Cirrhosis

It is known that toxoplasmosis rarely leads to various liver pathologies, most common of which is granulomatose hepatitis in patients having normal immune systems. Patients who have cirrhosis of the liver are subject to a variety of cellular as well as immunity disorders. Therefore, it may be considered that toxoplasmosis can cause more frequent and more severe diseases in patients with cirrhosis and is capable of changing the course of the disease. Cirrhotic patients are likely to form a toxoplasma risk group.

Toxoplasmosis and Lung Disease. Pulmonary toxoplasmosis is relatively rare in otherwise healthy subjects. In one case, a 41-year-old previously healthy male patient who presented to the emergency department of a hospital with a life-threatening case of pneumonia due to Toxoplasma gondii infection, which responded to specific therapy.

Clinical and image-based findings overlap with those for atypical pneumonias, and toxoplasmosis should be considered in the differential diagnosis—especially if immunoglobulin M-specific antibodies are detected.

Toxoplasmosis Can Mimic SLE

Toxoplasmosis may present with clinical signs usually characteristic of other diseases. A 48-year-old woman with a recent diagnosis of SLE was admitted to the hospital because of a fever, confused state, and convulsive episode. Her symptoms were interpreted as being compatible with lupus cerebritis.

Treatment with methylprednisolone resulted in a temporary improvement in the patient's condition. Nevertheless, during the next few weeks, her physical and mental condition deteriorated, and she died of massive pulmonary emboli. An autopsy revealed no signs of lupus cerebritis; however, disseminated cerebral toxoplasmosis was found. Cerebral toxoplasmosis is a rare complication of SLE that may be misdiagnosed as lupus cerebritis.

Toxoplasmosis and Neuroretinitis

Neuroretinitis is an eye disease usually seen in young healthy adults, that is characterized by rapid profound loss of vision and includes optic nerve head swelling, splinter hemorrhages, and variable eye inflammation. There are numerous causes that can cause a picture of neuroretinitis ranging from vascular to infectious to autoimmune. Patients with neuroretinitis caused by Toxoplasma infection respond well to treatment with systemic antibiotics and corticosteroids. Visual acuity may return in the range of 60-to 100 percent. Infectious causes of neuroretinitis, including toxoplasmosis, should be kept in mind in order to maintain visual acuity by early diagnosis and appropriate therapy.

Toxoplasmosis and Increased Risk of Traffic Accidents

The parasite Toxoplasma gondii infects 30-60% of humans worldwide. Latent toxoplasmosis, i.e., the life-long presence of Toxoplasma cysts in neural and muscular tissues, leads to prolongation of reaction times in infected individuals. It is not know, however, whether the changes observed in laboratory influence the performance of subjects in real-life situations. Researchers analyzed the presence of hidden toxoplasmosis in people involved in traffic accidents (total 146 accidents) and in the general population living in the same area (total 446 individuals) and found that relative risk of traffic accidents decreases with the duration of infection. These results suggest that 'asymptomatic' (having no signs) acquired toxoplasmosis might in fact represent a serious and highly underestimated public health as well as economic problem.

Diagnosis of Toxoplasmosis

Enlargement of lymph nodes of the neck, slightly elevated body temperature and discomfort are symptoms characteristic of many illnesses. One of these can be toxoplasmosis. Sometimes toxoplasmosis may be last to be recognized.

In many cases absence of specific additional examination guidelines can contribute to several problems with correct diagnosis. At the present time, the most reliable sample analysis methods are the examination of levels of antibodies IgG and IgM, and the histopathological verification. The authors also indicate that varying therapeutic effects using prophilactic treatment and insufficient additional examination could lead to diagnostic problems.

Prevention of Toxoplasmosis

In utero infection with Toxoplasma gondii may result in congenital defects such as hydrocephalus, chorioretinitis and mental retardation; these defects may be present at birth or may develop later in life. Prevention of this disease can be achieved in different ways.

- The most effective measure is to prevent the acquisition of the disease during pregnancy by avoiding risk factors for Toxoplasma gondii infection. Health education may decrease the incidence of toxoplasmosis during pregnancy by 60 %.
- Serologic screening during pregnancy is a preventive measure is based on to identify infected women. Treatment during pregnancy results in a significant reduction in the incidence of sequelae including severe handicaps. A third possible intervention is treating infected neonates.

Antibiotic treatment of infected children has a beneficial effect on the development of sequelae and the sooner therapy is started after birth, the better the outcome. This overview presents the potential benefits and harms of these different options available for the prevention of congenital toxoplasmosis .

How to Treat Ich Diseased Fish with Formalin

Formaldehype is most effective at ridding fish of common parasitic protozoan Black Spot, Clownfish, and White Spot marine ich infestations, as well as flukes, lice, and fungal diseases that both fresh and saltwater fishes can contract.

Although a toxic chemical that is hazardous in its pure form, by purchasing and using a standard over-the-counter "formalin" product, typically a 37% solution of formaldehype diluted with water (compare prices), it's rather easy to treat ich diseased fish.

Diagnosing Parasites on Marine Fish

- fish diseases
- parasites

Saltwater Aquariums Ads

- Fish
- Parasites
- Marine Fish Tank
- Koi Disease
- Marine Aquarium

Anyone who has had a saltwater aquarium for any length of time has probably experienced a parasitic outbreak. Since the treatment for each of the parasites is different, the biggest problem is usually trying to diagnose which of the three common parasites (Ich, Marine Velvet and Anemonefish Disease) are present on the fish and in the tank.

Cryptocaryon (A.K.A. Marine White Spot Disease, Ich)

Appears as white spots (sometimes described as appearing like salt granules) on the body and fins of the fish. These are more easily seen from above with a flashlight in a darkened room. When infected with Ich, the fish will usually respire heavily, appear restless, scraping against the substrate and rocks.

Amyloodiniosis (A.K.A. Oodinium, Marine Velvet or Coral Reef Fish Disease)

The fish will usually respire heavily, (Marine Velvet Disease usually attacks the gills first), appear sluggish, scraping against the substrate and rocks. The cysts are somewhat smaller than with Ich, appearing as a dust on the skin of the fish. As with Ich, Marine Velvet is more easily seen from above with a flashlight in a darkened room. The color of the fish may fade and the cysts may be visible on the head, body and fins.

Brooklynellosis (A.K.A Brooklynella or Anemonefish Disease)

Anemonefish Disease infects both the gills and the skin of the fish. One of the most distinct symptoms of this infection is the sloughing of the skin and increased mucus secretion. Infected fish will display rapid gilling, areas of discolored skin and stop feeding. It should be noted that Anemonefish are not the only fish which can get this disease. In fact, Seahorses will also contract it. This infection most frequently occurs in fish which have been recently shipped.

Bibliography

Andrews C.: *The Manual of Fish Health*, Voyageur Press, Stillwater, MN, 1999.

Axelrod HR and Untergasser D.: *Handbook of Fish Diseases*, T.F.H. Publications, Neptune, NJ, 1989.

Bennett, M. V. L.: *The Central Nervous System and Fish Behavior*, University of Chicago Press, Chicago, 1968.

Bruno D.W., Poppe T.T., *A Color Atlas of Salmonid Diseases*, Academic Press, London, 1996.

Burgess, Peter: *Tropical Fishlopaedia: A Complete Guide to Tropical Fish Care*, Howell Books, NY, 2000.

Chhapgar, B. F.: *Fishes of India*, Oxford University Press, Delhi, 2008.

Dawes, John: *Tropical Aquarium Fish: A Step-by-Step Guide to Setting up and Maintaining a Freshwater or Marine Aquarium*, New Holland Publishers, Holland, 2000.

Dholakia, A.D.: *Fisheries and Aquatic Resources of India*, Daya, Delhi, 2004.

Dubey, Pushpalata: *Fish Management and Aquatic Environment*, Daya, Delhi, 2006.

Emygdio, L. Cadima: *Fish Stock Assessment Manual*, Daya, Delhi, 2007.

Exell A, Burgess PH and Bailey MT: *A-Z of Tropical Fish Diseases and Health Problems*, Howell Book House, New York, 2001.

Ferguson, H.W.: *Systemic Pathology of Fish*, Iowa State Press, Ames, Iowa, 1989.

Frank, C. Edminster: *Fish Ponds for the Farm*, Agrobios, Delhi, 2004.

Grizzle J.M.: *Anatomy and Histology of the Channel Catfish*, Auburn Printing Co, 1976.

Hasler, A. D.: *Orientation and Fish Migration, in Hoar, W. S., and Randall, D. J.,* New York, Academic Press, 1971.

Helfman G., Collette B., & Facey D.: *The Diversity of Fishes*, Blackwell Publishing, UK, 2004.

Ingles, Lloyd G.: *Mammals of the Pacific States,* Stanford University Press. 1965

Ivantsoff, W.: *Natural History of Dalhousie Springs.* W. F. South Australian Museum, Adelaide. 1989.

Kishore, Braj Prasad Singh: *Fish Genetic and Biotechnology,* Swastik, Delhi, 2008.

Kumar, Arvind: *Fish Biology,* APH, Delhi, 2005.

Lewbart G.A. *Self-Assesment Color Review of Ornamental Fish,* Iowa State Press,1998.

Malvee, Sangeeta: *Fish Genetics,* SBS Pub, Delhi, 2008.

Martin, A.M.: *Fisheries Processing: Biotechnological Applications,* Chapman and Hall, Delhi, 2009.

Mehta, Varun: *Fisheries and Aquaculture Biotechnology,* Campus Books International, Delhi, 2006.

Nelson, J. S.: *Fishes of the World,* John Wiley & Sons, Inc.. , NY, 2006.

Petr, Tomi: *Fisheries in Irrigation Systems of Arid Asia,* Daya, Delhi, 2007.

Ponder, W. F. *Natural History of Dalhousie Springs.* South Australian Museum, Adelaide. 1989.

Ponniah, A.G.: *Fish Biodiversity of India,* NBFGR, 2002.

Reichenbach-Klinke H. H.: *Fish Pathology,* T.F.H. Publications, Inc. Neptune City, NJ. 1973.

Robertson, D.R.: *Fishes of the Tropical Eastern Pacific,* University of Hawaii Press, Honolulu. 1994.

Sandford, Gina: *Aquarium Owner's Guide,* DK Publishing, London, 2000.

Selvamani, B.R. and R.K. Mahadevan: *Fish and Fishery Culture: Assessment and Evaluation,* Campus Books International, Delhi, 2008.

Sharma, L.L. and A.K. Mathur: *Hand Book of Freshwater Ornamental Fishes,* Yash Pub house, Delhi, 2006.

Silva, De: *Fish Nutrition in Aquaculture,* Springer, Delhi, 2001.

Smith, R.I. and J.T. Carlton: *Lights' Manual: Intertidal Invertebrates of the Central California Coast,* University of California Press, Berkeley. 1975.

Trevor, A. Anderson: *Fish Nutrition in Aquaculture,* Springer, Delhi, 2009.

Ward, Barbara E.: *Varieties of the Conscious Model: Fishermen of South China,* Tavistock, London, 1965.

Warner, William W.: *Beautiful Swimmers: Watermen, Crabs and the Chesapeake Bay,* Little, Brown, Boston, 1976.

Index

L

M

N

O

P

R

S

T

U

V

W

Y

Z

❑❑❑